Space Ecology

Patrizia Caraveo

Space Ecology

From Earth to Moon and Mars

 Springer

Patrizia Caraveo
INAF
Milano, Italy

ISBN 978-3-031-78343-2   ISBN 978-3-031-78344-9   (eBook)
https://doi.org/10.1007/978-3-031-78344-9

The original submitted manuscript has been translated into English. The translation was done using artificial intelligence. A subsequent revision was performed by the author(s) to further refine the work and to ensure that the translation is appropriate concerning content and scientific correctness. It may, however, read stylistically different from a conventional translation.

Translation from the Italian language edition: "Ecologia spaziale - Dalla Terra alla Luna a Marte" by Patrizia Caraveo, © Author 2024. Published by Hoepli Editore. All Rights Reserved.

Editorial Contact: Marina Forlizzi
This Springer imprint is published by the registered company Springer Nature Switzerland AG
The registered company address is: Gewerbestrasse 11, 6330 Cham, Switzerland

If disposing of this product, please recycle the paper.

Contents

Ecology and Space Are Deeply Intertwined

Our idea of ecology is undoubtedly linked to the defense and preservation of the environment that surrounds us, an environment that is composed of goods that belong to someone and goods that belong to everyone because they are common goods. We know very well that human activity interferes with local and global ecological balances and this is particularly true for common goods, those that Roman law called *res communes* because by their nature they cannot be privatized, like the atmosphere or the oceans. Since they belong to no one, yet are available to everyone, the res communes, which in Anglo-Saxon law have become the *global commons*, are particularly exposed to the consequences of excessive exploitation. Since you don't have to pay to fish in the high seas, for example, our oceans are severely affected by overfishing. Unfortunately, the same happens for everything that is available for free, such as the collection of sand from river banks or extraction of water from the oldest aquifers, which were filled in geological eras but are being emptied at an alarming rate. It is the tragedy of the commons, a sadly familiar picture that must be fought precisely thanks to the growing ecological awareness. If we want to

P. Caraveo, *Space Ecology*, https://doi.org/10.1007/978-3-031-78344-9_1

leave a healthy planet to future generations, we need to promote a conscious and sustainable use of Earth's resources, but how far should we go? In other words, what are the boundaries of the environment we want to protect? Where does end what surrounds us and allows us to live and operate?

Human beings have been carrying out activities in space since 1957 and it was traditionally assumed that space activities were confined to space. Today the situation has radically changed: the quality of our life depends on the ability to use services offered by satellites and, to meet public demand, the number of satellites is literally exploding. We have realized that near-Earth space is a common good of humanity. Its dimensions are large, but not infinite. We are talking about a precious environment that needs to be preserved by learning to use it in a conscious and sustainable way. Above all, we must avoid it becoming a new example of the tragedy of the commons.

By generalizing the concept of ecosystem to include also the near-Earth space where satellites orbit, we are completing a long journey that began over 60 years ago, when one of the most important photos of our history, showing us the beauty and fragility of our planet was snapped on the spur of a moment.

Space That Inspires

It's the evening of December 24, 1968 and the astronauts of the Apollo 8 mission are the first human beings to fly over the hidden hemisphere of our satellite. The crew is made up of commander Frank Borman, on his second flight, command module pilot James Lowell, on his third flight, and lunar module pilot William Anders on his first flight. After spending the three days of the journey always in sight of Earth, the astronauts, who are tasked with photographing the surface of the Moon, feel isolated. For this reason, when the motion of their spacecraft takes them out of the shadow of the Moon, they greet with joyful surprise the sight of the Earth's hemisphere illuminated by the Sun. The transcript of what they say to each other shows this very clearly

Anders: Oh my God! Look at that picture over there. Here's the Earth coming up. Wow, that's pretty.
Borman: Hey, don't take that, it's not scheduled.
Anders: [laughter] "You got a color film, Jim? Hand me that roll of color quick, would you…
Lovell: "Oh man, that's great."

Anders is the first to be struck by the sparkling Earth. Commander Borman, who is tasked with taking as many photos as possible of the lunar surface to improve the cartography that will be used to choose the landing sites for future Apollo missions, attempts in vain to dissuade him from wasting time on an unplanned activity. Anders laughs, asks Jim Lowell to hurry up and pass him the camera with the color film, and snap the photo of the Earth rising. NASA will call it Earthrise and I think it should be dedicated to Bill Anders who, at ninety, left us on June 7th 2024 crashing with his plane. In addition to being included in the list of the 100 photos that changed the world, compiled by Life magazine in 2003, the rising Earth had the honor of the cover of that issue, 35 years after the Apollo 8 mission.

I am never tired of admiring the beauty of the Earth shining in the black sky with the white of the clouds and ice and the blue of the oceans against the almost uniform gray of the moon (Fig. 1).

Fig. 1   Earthrise- The rising earth. (Credit NASA)

Two celestial bodies that appear very different but have a common origin. As hard as it may be to believe looking at this image, the Moon is a piece of Earth. We know it for sure on the basis of the analysis of lunar soil samples collected by subsequent missions, which brought 12 American astronauts to walk on its surface. The exploration of our satellite has had very important scientific and technological results, nevertheless all the lunar astronauts agreed that what struck them most was seeing the Earth so bright and so isolated in the darkness of space. Eugene Cernan, commander of the Apollo 17 mission, declared "We went to explore the Moon, and in fact discovered the Earth".

Many people think, probably with reason, that this photo, in its explicit simplicity, contributed to the development of ecological sensitivity that had a turning point thanks to the publication of Rachel Carson's bestseller "Silent Spring". With over half a million copies sold in 24 countries, the book increased public awareness and concern for living organisms, the environment, and the inextricable links between pollution and public health.

Indeed, half a century ago, global warming was not yet a subject of general interest, but the problem of air and water pollution was already widely debated, even if traditional America was largely unaware of environmental issues and how a polluted environment is harmful to human health.

Earthrise is an image that moves consciences and certainly makes you want to defend our planet, vast areas of which are polluted by industrial activities, car emissions, indiscriminate discharges. This may be the *raison d'être* of the first Earth Day, organized on April 22, 1970 under the impetus of Democratic Senator Gaylord Nelson who thought of giving voice to the concern of young generations for the state of the environment. April 22 is not an anniversary but rather a convenient date to allow university students to participate in the demonstrations that had been organized everywhere in the United States. The public responded en masse: twenty million people were counted, a huge figure, corresponding to 10% of the American population of the time. A clear signal for US politics which, in the following years, approved laws for the control of air and water quality and for the protection of endangered animal species. For the first 20 years, the *Earth Day* was just an American initiative, then, in 1990, it became global involving 200 million people in 141 countries. Since then, it has never stopped growing, focusing its attention on climate change, as well as on the protection of the environment.

I would like to conclude this paragraph dedicated to how space has changed our perception of our planet with another iconic photo of the Earth taken by the Voyager 1 probe on February 14, 1990, more than 12 years after the start of its journey. Before the onboard cameras were turned off to save energy, NASA had accepted Carl Sagan's proposal to turn the probe towards the inner solar system to take a family portrait of the planets (Fig. 2).

The result of those 60 shots is one of the most evocative scientific space image of all time. Our planet appears as a barely perceptible pale blue dot in a streak of light produced by the reflection of sunlight.

Four years later, in 1994, Sagan published the book "Pale Blue Dot: A Vision of the Human Future in Space", where we find a profound reflection on that photograph

Look again at that dot. That's here. That's home. That's us. On it everyone you love, everyone you know, everyone you ever heard of, every human being who ever was, lived out their lives. The aggregate of our joy and suffering, thousands of confident religions, ideologies, and economic doctrines, every hunter and forager, every hero and coward, every creator and destroyer of civilization, every king and peasant, every young couple in love,

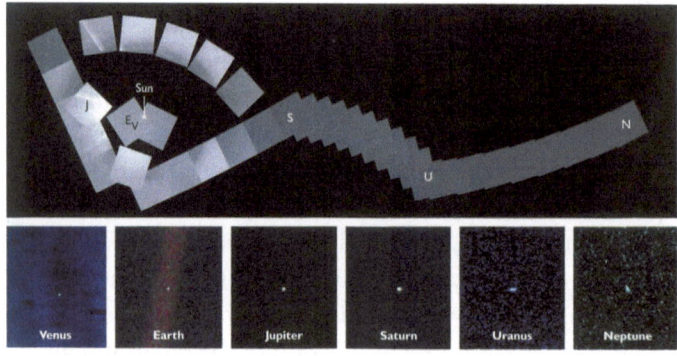

Fig. 2   Family portrait of the solar system taken by the Voyager 1 probe. (Credit NASA)

every mother and father, hopeful child, inventor and explorer, every teacher of morals, every corrupt politician, every "superstar," every "supreme leader," every saint and sinner in the history of our species lived there--on a mote of dust suspended in a sunbeam.

The Earth is a very small stage in a vast cosmic arena. Think of the rivers of blood spilled by all those generals and emperors so that, in glory and triumph, they could become the momentary masters of a fraction of a dot. Think of the endless cruelties visited by the inhabitants of one corner of this pixel on the scarcely distinguishable inhabitants of some other corner, how frequent their misunderstandings, how eager they are to kill one another, how fervent their hatreds.

Our posturings, our imagined self-importance, the delusion that we have some privileged position in the Universe, are challenged by this point of pale light. Our planet is a lonely speck in the great enveloping cosmic dark. In our obscurity, in all this vastness, there is no hint that help will come from elsewhere to save us from ourselves.

The Earth is the only world known so far to harbor life. There is nowhere else, at least in the near future, to which our species could migrate. Visit, yes. Settle, not yet. Like it or not, for the moment the Earth is where we make our stand.

It has been said that astronomy is a humbling and character-building experience. There is perhaps no better demonstration of the folly of human conceits than this distant image of our tiny world. To me, it underscores our responsibility to deal more kindly with one another, and to preserve and cherish the pale blue dot, the only home we've ever known.

In February 2020, to commemorate the 30th anniversary of the historic image, NASA published a revisitation of the Pale blue dot, which retains all its evocative power (Fig. 3).

Fig. 3 Pale blue dot. (Credit NASA)

Space That Controls

Apart from inspiring our ecological sensitivity, satellites are our global eye for monitoring the health status of the planet and for identifying the activities or industrial realities that pollute the most. For this meticulous control work, a single satellite alone, no matter how high-performing, is not enough. It is necessary to exploit all the wealth of data that has been accumulated over the years and continues to grow. For this reason, a non-profit coalition called Climate Trace (an acronym for Tracking Real-time Atmospheric Carbon Emission) was born. Exploiting the power of artificial intelligence, Climate Trace combines the data collected from 300 satellites and 11,100 sensors scattered at sea, on land and in the air. In doing so, it has identified over 70,000 individual sources of greenhouse gases. Their numbers are independent of what governments declare and Climate Trace claims that their results are even more accurate because they are evaluated locally. Not that the task is always very easy, in fact often the measures are not direct and it is necessary to use proxies. For steel mills, for example, satellite information on the heat released from furnaces is used to estimate the steel produced. While for power plants that produce electricity by burning fossil fuels, satellite measurements of the steam released from chimneys are used.

Perhaps for this reason in the ranking of polluters the first positions are all held by the extractive industry *oil and gas*. Fossil fuels, which produce CO_2 when burned, begin to release greenhouse gases when they are extracted because oil deposits always contain "methane bubbles" that are released (often burned) during extraction. The methane released along with the procedure of "flaming", i.e. the flames that burn continuously fueled by gas, give to the *oil fields* the title of activities that produce the greatest amount of greenhouse gases. In fact, the production of greenhouse gases from oil extraction is a great example of the usefulness of Climate Trace, since the quantities reported on their site are triple w.r.t. those declared to the UN by companies that, perhaps thinking about a probable carbon tax, declare the smallest possible amount.

But extraction is the first step, then the oil must be transported (almost always by ship) to the places where it is refined and then used, not without having moved further in the distribution process. Refineries are another of the sectors with high production of greenhouse gases. Significant contributions also come from cement plants and steel mills.

But, on the Climate Trace map, there are also natural producers of methane such as landfills and large rice fields.

Quantifying the who, where and how much is particularly important when starting to discuss the thorny issue of compensations that rich countries (which produce large amounts of greenhouse gases) should provide to the poorest ones that are hit more dramatically by global warming despite having contributed very little to the emission of greenhouse gases. It is the chapter of *loss and damage*, a principle on which rich countries say they agree, but on which they do not want to commit because they fear signing a blank check.

However, it is clear that if we are talking about reducing greenhouse gases by a certain date, we need to be very clear about what the starting level is. Therefore, since the rule *"you can manage only what you can measure"* applies, it is essential to have precise measurements.

Space That Offers Services

When you make a payment, a bank transfer, an investment, when you check the weather forecast, when you are guided to your destination by geolocation apps, when you follow a sport event that is taking place on the other side of the planet, you are using services offered by satellites.

Without satellites, the world's economy would stop, banks could not operate because every movement of money, large or small, has a *time tag* provided by GPS satellites, planes and ships would have some trouble finding the right routes and we would be late for appointments since we all rely on geolocation. No satellites, no accurate weather forecasts, no worldwide broadcasts, no control of the territory to prevent landslides and floods or to provide aid to people affected by natural disasters.

Different services imply different satellites, of different sizes, describing different orbits and that can be used in different ways. We go from public utility services, financed by space agencies, defense ministries or national or international institutions, which can be used for free, to services customers should pay for, mostly connected to earth observations (which can have civil or military purposes), to telecommunications and internet services.

While agencies and governments invest in public utility programs, private entrepreneurs seek profit by offering services that may interest a large audience such as satellite TV, high-quality observations for precision agriculture but also for borders control and, very recently, global internet connections.

The satellites that operate day and night above our heads are also essential to achieve at least half of the 17 Sustainable Development Goals that the United Nations launched in 2015 by 2030.

Thus, in addition to inspiring and controlling, space promotes the growth of our society.

Even if we are not always aware of it, our way of life is intimately connected to the services offered by a plethora of satellites orbiting the Earth. We do not see them but the satellites constitute an infrastructure of our society, a complex infrastructure that has required, and continues to require, substantial public and private investments but that, unfortunately, does have a number of environmental impacts we should be aware of.

Circumterrestrial Space and Satellites

Circumterrestrial space extends from a height of 100 km above the surface (known as the Karman limit) to the geostationary (or geosynchronous equatorial) orbit (GEO) which is 35,786 km above the equator. Only the geostationary orbit is required to be precisely equatorial and precisely at that height, all other types of orbits are free in terms of height and inclination.

Within this space, satellites are classified into four major types, based on their orbital characteristics:

LEO (Low Earth Orbit) satellites orbit between 200 and 2000 km above the Earth's surface, crossing (and interacting) with the highest layers of our atmosphere. We are talking about the thermosphere (which goes from 200 to 500 km in height) and the exosphere which extends up to 2000 km, even if it is difficult to define a true limit. Although the density of the atmosphere decreases with increasing height, it is still a viscous medium and its friction influences the motion of satellites, whose orbits are modified. The lower their altitude, the more their orbit height decreases over time, until the heat generated by

© The Author(s), under exclusive license to Springer Nature Switzerland AG 2025
P. Caraveo, *Space Ecology*, https://doi.org/10.1007/978-3-031-78344-9_2

friction with the increasingly dense atmosphere destroys them. For this reason, the operational life of a satellite in low orbit is intimately connected with the higher layers of the atmosphere, which provide a natural mechanism to clean the orbits, a mechanism that becomes more efficient when enhanced solar activity causes the atmosphere to "swell". If you want to keep a satellite in low orbit operational, orbital decay must be controlled with orbit-raising maneuvers, as it is periodically done with the International Space Station, which is pushed to a higher altitude twice a year (Fig. 1).

Higher up, outside the atmosphere, we find the Medium Earth Orbit (MEO) which operate between 8000 and 20,000 km above the Earth's surface.

Then, precisely above the equator at a height of 35,786 (known as Clarke's orbit from the name of the great science fiction writer who first used it in his stories), there are geostationary satellites that always appear in the same position because their orbital period is identical to Earth's rotation.

Finally, we find highly elliptical orbits that at apogee (the farthest point) go well beyond the geostationary orbit and reach heights of over 100,000 km, like the INTEGRAL satellite of the European Space Agency. These are orbits that allow instruments to operate for days without ever having the Earth in their field of view.

Each orbit has its own travel time that depends on its height. Low orbits, roughly at 500 km, are covered in 90 min, then the period to described a full orbit lengthens with altitude up to 24 h for the geostationary orbit, and even several days for very elliptical orbits. A rendering of all the objects orbiting Earth at different height id shown in Fig. 2.

The orbit described by a satellite is chosen on the basis of its scientific, social, military or commercial purpose, from

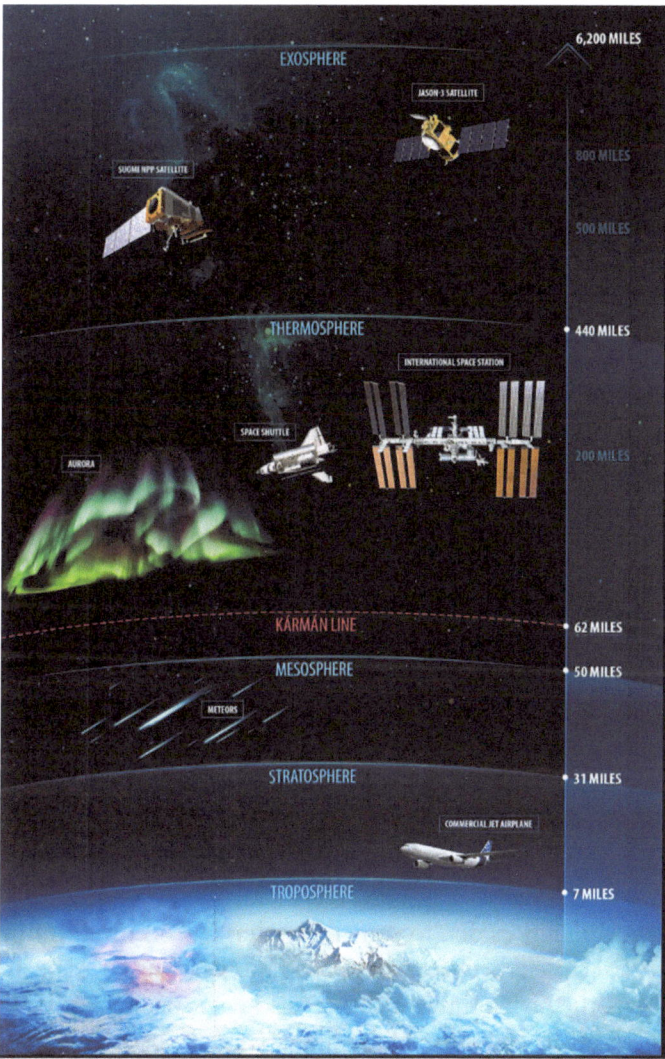

Fig. 1   diagram of the characteristics of the atmosphere as a function of height above sea level. (Credit NOAA)

Fig. 2 Rendering of circumterrestrial space showing a clear crowding in low orbits. (Credit ESA)

where it is launched and the geographical position of the ground stations. Earth observation satellites, for example, tend to have low "polar" orbits, that is, very inclined with respect to the equator, in order to fly over (and observe) strips of the globe from the south pole to the north pole. In the next orbit, the Earth (which rotates 15° per hour because it covers a full 360° turn in 24 h) will have rotated and the strip flown over will be shifted compared to the previous one (Fig. 3) allowing the satellite to cover the entire globe in a few days after which, it will repeat the passages over the areas already observed making it possible to capture differences due to something that has occurred in the time elapsed between the observations (revisit time). To lower the revisit times, a constellation of satellites is needed in order to cover the globe like a web. The same applies, even more so, for constellations of satellites that want to offer the service of global internet connection and must therefore be able to offer continuous connection.

As already mentioned, the International Space Station is a huge structure in low orbit whose inclination is the result

Fig. 3 Passages of the SENTINEL 2 satellite from the European Commission's Copernicus program. (Credit ESA)

of an international compromise between NASA and ROSCOSMOS (the Russian space agency) which have launch bases in different locations. While NASA launches from the Kennedy Space Center, in Florida (latitude 28.5°), ROSCOSMOS launches from Baikonur in Kazakhstan (latitude 51.6°). Since a satellite automatically acquires an orbit with an inclination equal to the latitude of the launch base, it was clearly necessary to choose the most convenient one. Since orbital mechanics says that it is easier to increase the inclination of an orbit rather than decrease it, the ISS has the Kazakh orbit and NASA must correct the inclination of what is launched from Florida.

The same considerations of orbital mechanics make us understand that, to launch satellites in equatorial orbit, having a launch base near the Equator is extremely important. This consideration guided the Italian choice to build a base off Malindi in Kenya. Thanks to a visionary idea by Luigi Broglio, air force general and professor of aerospace engineering at the University of Rome, in the mid-60s, two oil platforms were converted into launch platforms for small satellites in low Earth orbit. In total, between 1967 and 1988, 27 launches were carried out. In addition to the orbiting of Italian satellites, the equatorial launch base was used by NASA for the launch of its three scientific satellites of the SAS series (Small Astronomy Satellite) which wanted to take advantage of the benefits offered by the low equatorial orbit, which is the least disturbed by cosmic rays. The first SAS, known as Uhuru (freedom in Swahili), because it was launched on December 12, 1970 on the anniversary of Kenya's independence, opened the window of astronomy in the field of X-rays and earned its creator, Riccardo Giacconi, the Nobel Prize for Physics in 2002.

But in addition to offering advantages for scientific satellites in low orbit, let's not forget that the geostationary orbit, which is the most commercially attractive, must also be equatorial. Thus, it is clear why the European Space Agency (ESA) made a far-sighted choice in building its spaceport in French Guiana which, thanks to the Ariane launchers, has dominated the market for geostationary launches for decades, until the revolution of Space Economy.

Geostationary orbits are used for telecommunications and weather satellites (which can cover vast areas of the globe). The European Meteosat satellites, for example, are stationed on the equator above Africa to cover all of Europe. Since height and inclination are fixed, the geostationary orbit can accommodate a limited number of satellites that must be positioned with accuracy to avoid collisions and interference between their transmissions. Moreover, in the case of telecommunications satellites that must act as a radio bridge with the television stations of a certain nation, it is necessary for the satellite to be positioned on the equator above or in view of a certain geographical area (Fig. 4). This implies that the requests for geostationary spaces are not uniformly distributed on the equator. While there are many contenders for the most populated and industrialized areas

Fig. 4 Rendering of the geostationary orbit with the aligned satellites. (Credit ESA)

corresponding to the American continent and Eurasia, the uninhabited areas, for example above the Pacific, are empty because no one requests them. All this has created the need to control geostationary slots at an international level. This important task is carried out by the International Telecommunication Union (ITU), a United Nations agency that regulates the global use of radio frequencies. In the case of satellites, where each one must be able to operate without creating harmful interference with all the others, the ITU assigns the transmission and reception frequencies for each transmitting and/or receiving satellite in each orbital category. All this information is then reported in the Master International Frequency Register (MIFR).

For this reason, anyone who wants to launch and operate a satellite must first ask ITU for authorization to use the transmission and reception frequencies (which must be different otherwise the satellite would interfere with itself). Once the frequencies are obtained, it is necessary to use them within a certain period of time, otherwise the authorization to use this immaterial but very precious good will expire. The hunger for frequencies is such that the ITU has requests for frequencies for a million potential satellites, most of which will never be realized.

Unlike the authorization to use frequencies, which is regulated globally, launch authorizations are managed nationally because there is no global authority to dictate the rules on the use of orbits, on the number of satellites that can be launched, on the characteristics they must have, on how to dispose of those no longer working, just to mention the problems we are seeing arise around the Earth. Everything that is launched from the US territory, for example, must go through the Federal Communications Committee, a federal body that, having made the necessary evaluations, issues, or not, the launch authorization. The international treaties (see appendix) only require that satellites be registered. In this way, however, the situation is monitored but not regulated.

While the assignment of frequencies is a common and necessary step for all satellites, in the case of those that must operate from the geostationary orbit the ITU also assigns the position which is calculated to be at least 9 km away from neighboring satellites. Also the assignment of the position is not indefinite: the company that manages the GEO satellite must commit to freeing the place at the end of the orbital life by moving the satellite to a graveyard orbit, where it will not cause disturbance.

The Circumterrestrial Space Needs Rules

From this brief overview it is clear that the circumterrestrial space does not enjoy uniform regulation. While the management of frequencies used by satellites is regulated globally by a United Nations body, the occupation of orbits, except for the geostationary one, is left to national bodies that act independently of each other and, with every probability, follow different practices. What can be authorized for launch in one nation might not be in another that has adopted stricter rules on orbit occupation or environmental impact.

Radio frequencies and circumterrestrial orbits are clear examples of common goods that are perceived and managed in very different ways.

While it is clear that radio frequencies, although extending over a wide fraction of the electromagnetic spectrum, are a finite asset very requested and therefore need to be managed globally, circumterrestrial space is still seen as an unlimited resource freely available that can be used and occupied by those who have technical and financial resources to do so.

For this reason, many players feel the need for shared rules well beyond international treaties, developed to regulate activities in space over half a century ago when space was the prerogative only of a few great powers and it was necessary to avoid the onset of conflicts. It is not a coincidence that the 1967 treaty on the use of outer space (OST, reported in full in the Appendix) focuses especially on the prohibitions of using space for military purposes, of claiming sovereignty over the Moon or other celestial bodies that must be explored with the outmost respect for the environment. The treaty does not prohibit commercial activities but requires that national governments take responsibility for control.

Subsequent elaborations to the treaty require everyone's commitment to rescue astronauts in danger, to cover any damages caused by space probes and to provide the details of everything that is launched in order to have an always updated census of all the probes.

To date, there are as many as 91 Nations involved in space activities (whether institutional or commercial) mostly concentrated in the circumterrestrial orbit. In this changed landscape, it is clear that the management of circumterrestrial orbits falls into a legislative void since the subject is not at all covered by international treaties.

The important exception of the geostationary orbit, however, shows that, when the problem of occupying orbital slots which are available in limited number becomes crucial for commercial exploitation, it is possible to find solutions recognized by the actors even beyond the international treaties that require long and complex negotiations, and that do not always end successfully.

The Explosion of Numbers

Artificial satellites are certainly not a novelty of recent years. Humanity's space adventure began in October 1957 with the launch of Sputnik and, as time went by, space has entered with an ever increasing importance into our lives. There are very few human activities that do not use, at least in part, services offered by satellites.

It is the revolution of the Space Economy that has radically changed the space landscape by expanding both the number of service producers and that of users. While traditional players (civil space agencies and their military equivalents) continue their activity, now it is the private sector that plays the lion's share in all the most profitable activities such as earth observations and telecommunications.

To appreciate the magnitude of the change due to the forceful and pervasive entry of private entities into space activities, just look at the graph that shows the number of satellites launched each year. While for about half a century, roughly from the mid-60s to 2015, the number of satellites launched each year, albeit with variations, hovered around 200 units, after 2015 there is a marked increase that then becomes a dizzying growth so much so that, during 2023,

© The Author(s), under exclusive license to Springer Nature Switzerland AG 2025
P. Caraveo, *Space Ecology*, https://doi.org/10.1007/978-3-031-78344-9_3

about 2600 satellites were launched, over 2000 of which with an American flag (Fig. 1). While it is clear that above our heads something of historical significance is happening, a question naturally arises: what is being launched? And for what use?

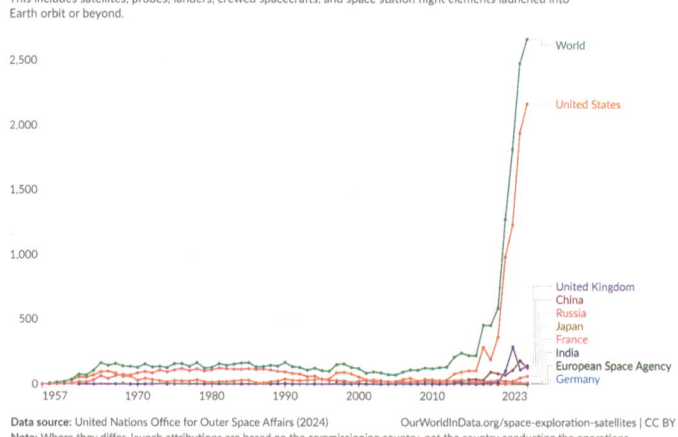

Fig. 1 Evolution of the number of objects launched over time. By objects, we mean satellites, probes, landers, crewed spacecraft, and flight elements of the space station launched into Earth orbit or beyond

These data are based on the national launch registers presented to the UN by the participating nations. According to UN estimates, the data capture about 88% of all objects launched

When an object is indicated by the source as launched by one Country on behalf of another, the launch is attributed to the latter Country. (Credit "Data Page: Annual number of objects launched into space", part of the following publication: Edouard Mathieu and Max Roser (2022)—"Space Exploration and Satellites". Data adapted from United Nations Office for Outer Space Affairs. Retrieved from https://ourworldindata.org/grapher/yearly-number-of-objects-launched-into-outer-space [online resource])

Looking at the census of active satellites in May 2023 https://nanoavionics.com/blog/how-many-satellites-are-in-space/ we can find the answers we are looking for.

Missions	Number of satellites
Telecommunications	3135
Earth observations	1052
Technological satellites	383
Navigation	154
Space science	108

Clearly, the categories with the highest number of satellites are those of telecommunications and earth observations. However, the same site tells us that as much as 72% of the 7650 active satellites at that time were small mass objects, weighing less than 600 kg. In fact, the class of small mass satellites is divided into at least three categories: starting from nanosatellites, with a weight between 1 and 10 kg, moving to microsatellites, with a weight from 10 to 200 kg, and ending with minisatellites, with a weight from 200 to 600 kg.

To be precise,

Category	Number of satellites
Minisatellites	2379
Microsatellites	331
Nanosatellites	790

The extraordinary growth of recent years is due almost entirely to the constellations of "small" satellites dedicated to earth observations and telecommunications. It is worth noting that only 34% of the satellites launched in the '90s fell into this category, while since 2020 "small" satellites represent 94% of all the satellites launched.

This change is clearly seen by comparing the mass distribution of satellites launched into low orbit with a height of less than 600 km before and after 2010 provided on J. McDowell's site.

While before 2010 the satellites launched were counted in tens and most of them had a mass of a few tons, after 2010 the numbers explode but these are only mini and nano satellites.

Nanosatellites, commonly known as cubesats, have changed the Earth commercial observations while minisatellites have ushered in the era of global internet connection.

Note that the figures give a cumulative view. The numbers do not represent the satellites launched each year but the total of satellites of that class active in orbit in a given year (Fig. 2).

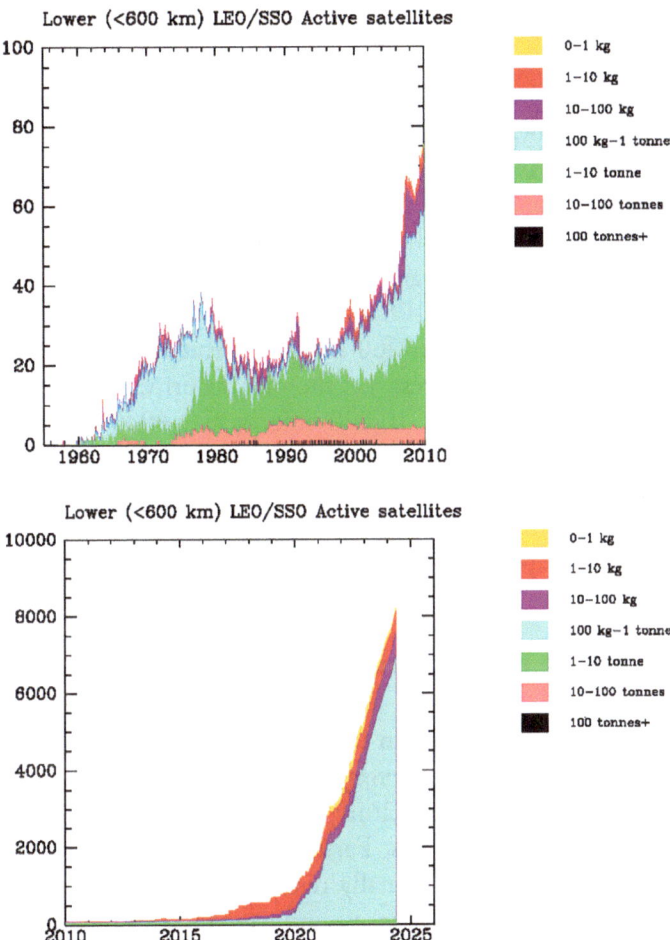

Fig. 2 Mass distribution of satellites launched until 2010 (above) and after 2010 (below) the blue part is dominated by the satellites of the Starlink constellation. (Credit site managed by J. McDowell)

The Updated Census

Considering that the number of satellites in orbit grows as time goes by, it is necessary to refer to the continuously updated census of the population of satellites in orbit that can be consulted on various dedicated sites. One of the most information-rich is that managed by J. McDowell https://planet4589.org/space/stats/active.html which, as of Feb. 2 2025, lists 11,108 active satellites, 550 of which are in geostationary orbit, 260 in MEO orbit and 23 in very elliptical orbits. The rest, i.e. the vast majority, are in LEO orbits that host about 10,000 objects, over 6930 of which are part of the Starlink constellation.

Dividing the satellites according to their mass, we see that about 9671 have a mass greater than 100 kg while those with mass under 100 kg are 1389.

Small Is Beautiful

Let's see what happens in the field that goes under the generic name of *remote sensing*, meaning observations from afar. Thanks to the miniaturization of instruments and the decrease in launch costs, Earth observations have evolved radically. While, traditionally, remote sensing satellites were (and are) large and expensive and were (and are) managed by space agencies (or by defense ministries), now there are many private companies that obtain and market the data. Observations are made by medium-small satellites, single or in constellations capable of providing almost continuous coverage of a certain region observing in optical and infrared, but also with radar instruments capable of piercing the clouds and operating in the dark.

There are many companies that provide images of the Earth aimed at land control, traffic monitoring, rivers and glaciers surveillance, precision agriculture and the myriad of services that we tend to take for granted without considering where they come from. We are talking about satellites of sizes ranging from cubesats of a few kilograms to large spacecraft in the ton class. Different masses mean vastly different costs but, curiously, the performance of the smaller satellites can be anything but negligible. In fact, cubesats were born as a gym to train new generations of aerospace engineers and have a relatively short history. They are named after a cube of 10 cm per side which represents the unit U. We therefore speak of satellites of 2,3,4,6 U depending on the size. The idea has always been to design a mission with a specific purpose and very low cost that uses instrumentation made from standard components easily available on the market. At first, cubesats did not seem interesting to space agencies, but private entrepreneurs have sniffed out their potential especially in the field of Earth observations because a constellation of cubesats, positioned judiciously, can provide continuous coverage of the planet unlike large satellites that have a time for "revisiting" of several days or weeks. Commercial interest means investments that turn into the growth of the number of cubesats launched each year (Fig. 3).

Planet Labs, for example, has a constellation of 190 *doves* of 3U that produce images with a resolution of a few meters for several hundred customers. Scientists have also realized the potential of cubesats that can be used for various purposes: from studying solar activity to monitoring variable X sources, to characterizing the circumterrestrial environment, to mapping animal migrations. Their use is not limited to low orbits: Capstone, the first mission of the Artemis

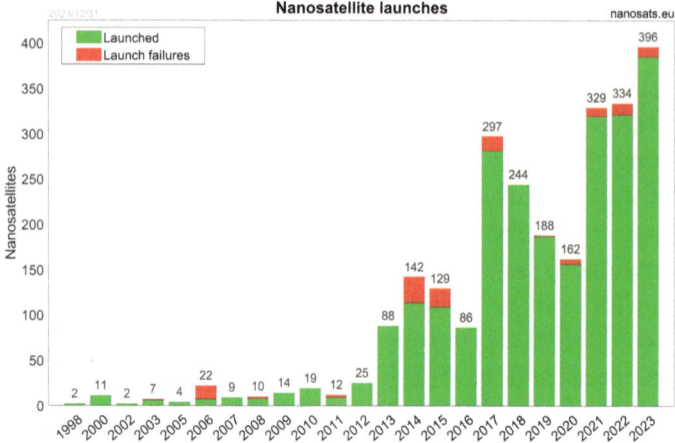

Fig. 3 Growth in the number of smaller satellites. (Credit nanosats.eu)

program for NASA's return to the Moon, is just over a cubesat, while Marco 1 and 2 have already arrived at Mars and the Italian LICIAcube has recorded the impact of the NASA DART mission against the asteroid Dimorphos in orbit around the larger Didymos.

Nothing can be hidden from the eye of satellites that record the damage of drought and floods, volcanic eruptions and sandstorms, ocean temperatures and wind speed, troop movements and the devastation of combat. Thanks to this new "transparency", images from space allow us to monitor climate change, to have increasingly accurate weather forecasts, but also to follow international crises and to verify the statements of one side and the other.

According to all indicators, earth observations are a lucrative business in continuous growth. All this without considering that, in case of a war, the data are also of strategic value.

In fact, the defense ministries of all nations, while continuing to invest in their own satellites for territory control, do not disdain to use private companies whose data are not

Fig. 4 The image of the damage from the MAXAR Technologies Twitter account

covered by military secrecy. The damage to the bridge in Crimea was recorded in the images of Maxar Technologies which is a supplier of the American government (Fig. 4).

But observations from above are also very valuable in case of more or less large environmental problems. When, on September 26, 2022, the pressure of the methane pipelines under the Baltic Sea suddenly dropped, it was the data from one of the *doves* of the Californian company Planet Lab that clarified what had happened, showing the gas bubbles gurgling from the pipe leaks (Fig. 5).

As already mentioned, the *doves* are launched in groups, or rather in "*flocks*", precisely to provide continuous coverage of the entire planet. *Doves* are small but numerous and it is therefore likely that some of them be at the right place at the right time. This rapid response is not possible for large satellites that are unique or form constellations of few components. In the days following the incident, data arrived from the Sentinel-2 satellites of the European Copernicus system and the American Landsat. But in addition to seeing the bubbles, it was necessary to measure the

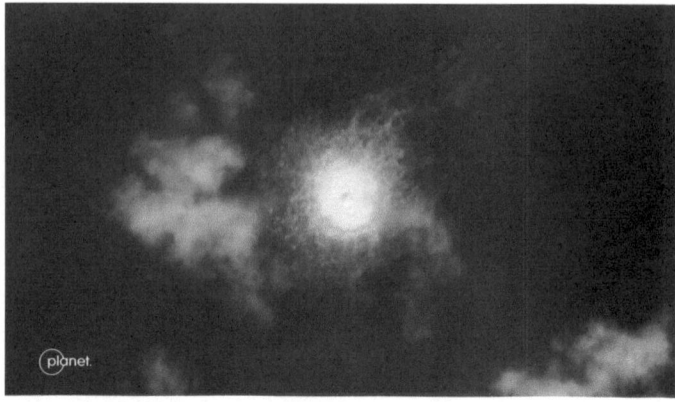

Fig. 5 The image, obtained on September 26, 2022 from one of the 200 Dove satellites managed by Planet Labs in San Francisco, shows the stain created by the gurgling methane. (Credit Planet Labs PBC)

amount of methane released. The two methane pipelines were not in operation but contained gas under pressure that continued to bubble for days. Measuring methane is not trivial, especially over water. On September 30, GHGSat, a company specialized in monitoring methane leaks, went into action. Using radar sensors, the methane leak was measured at 80 tons per hour (Fig. 6). Note that the gas had been bubbling for 4 days and the rate of leakage was certainly lower than the initial one. Taking into account these factors, a value was estimated that makes it the most significant methane leak ever recorded by satellites, probably greater than the loss of 40,000 tons of methane measured over the course of 17 days above an oil platform in the Gulf of Mexico in December 2021.

In addition to possible damage to aquatic life, methane is a greenhouse gas tens of times more efficient than carbon dioxide in trapping heat and so the extent of the environmental damage caused cannot be neglected.

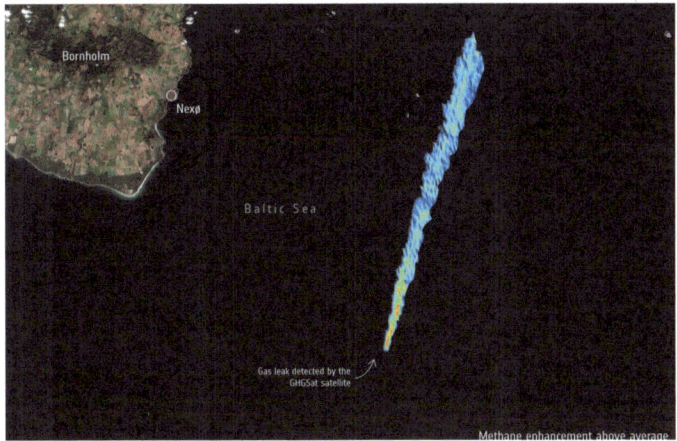

Fig. 6 The methane leak seen from a GHGSat satellite, on September 30. (Credit GHGSat and ESA)

However, as gigantic as the leak may appear, it is a drop in the ocean of methane that is released into the atmosphere over the course of a year. Experts say it is equivalent to a day and a half of methane losses from oil fields alone. A fact that reminds us that, during the COP26 in Glasgow in 2021, 120 nations signed the Global Methane Pledge in which they committed to reducing methane emissions by 30% by 2030. To achieve this goal, the oil industry must be persuaded not to release (or burn) unsold methane, introducing maximum allowed values with incentives if they are respected, or fines if they are exceeded.

The business of GHGSat, and other companies in the same sector, is precisely to monitor methane emissions to provide industries and international organizations with the necessary data.

Towards Global Connection

But satellites can do much more than observe. The world of telecommunications has been literally revolutionized by the use of satellites that makes it possible to access the internet from anywhere in the world. The signal is distributed by a dense network of satellites that orbit about 500 km high, in order to minimize transit time and provide quick responses to users. The forerunner of orbital internet is Elon Musk with his Starlink network, born to provide fast internet to over 10 million Americans and Canadians living in rural areas but now morphed into a global power. According to the census of https://planet4589.org/space/stats/acdec.html, as of Feb.2 2025, the Starlink constellation already has 6935 satellites in orbit and plans to soon have 12,000, with a possible extension of another 30,000. These are numbers that would have been unthinkable just a few years ago, instead SpaceX has surprised everyone with a launch rate that has made the company the absolute leader in the field of launchers. Fully exploiting its three ramps, scattered between California (at the Space Force base in Vandenberg) and Florida (at Cape Canaveral, where SpaceX rents platform 39A from NASA and 40 from the Space Force), the company can manage simultaneous launches from the two independent control rooms at the Hawthorne center (in California). The activity never stops. To get an idea of how frantic the launch sequence is, consider that in April 2024 there were 12 launches, 9 were devoted to Starlink and each brought into orbit 23 second-generation satellites, 1 was a multiple launch with several small satellites, 1 was military and 1 launched two satellites of the European Galileo constellation (which had to resort to the services of a private American company because there were no European launchers available). In May, the launches carried out by SpaceX were 14 and 11 of these carried Starlink satellites. One of the other 3 launches was purchased by the

European Space Agency for the Earthcare mission, while the other 2 were earth observation satellites for a US government agency and for a private company, always American, which has contracts with the Department of Defense.

Such a high number of launches is made possible by the reuse of the first stage of the rockets, complete with engines, which is quickly checked and prepared for the next takeoff. So far the *turnover* was as fast as 21 days and the number of reuses for a Falcon 9 has reached twenty-one (with the launch of May 17, 2024).

Such a remarkable results was achieved in just over 20 years, that is, since the visionary Musk decided that launchers should not be disposable any more, but rather recovered and recycled.

On this bet, which many experts judged almost impossible, he founded his extraordinary fortune in the Space Economy arena (Fig. 7).

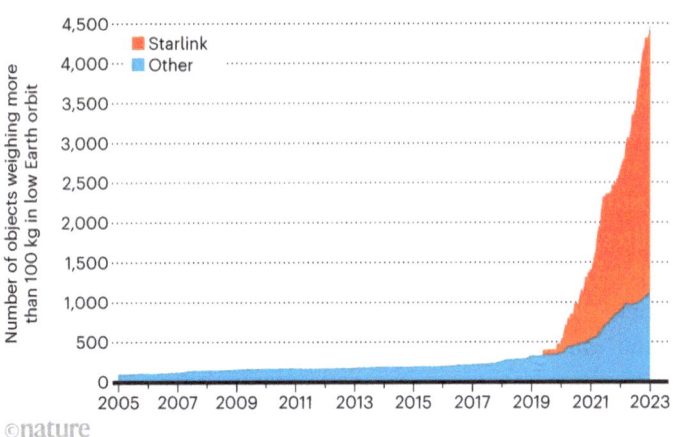

©nature

Fig. 7 Starlink satellites represent the majority of satellites launched since 2020 and, given their continuous growth, they are proportionally becoming more and more important. (Credit Nature). Indeed, as of Feb.2 2025, active Starlink satellites are 6935 while the entire satellites' populations in lower Earth orbits amounts to 9325 objects

He began launching his Falcon 9 rockets in June 2010 and now the count has reached over 290 with only two failures in the initial period and a problem with the second stage engine on July 11, 2024. The enquiry on the malfunction, requested by the Federal Aviation Administration, did momentarily stop the launches, which resumed on July 27.

In 2014 the first attempts at landing the first stage were performed, and after few failures, this daring maneuver became the standard for Falcon9 launches with continuous efforts to shorten the time between recovery and the next launch. This is one of the ingredients of SpaceX's absolute supremacy in the launcher market. To understand its dimensions, consider that during 2023 SpaceX has carried out 96 successful launches (in addition to the two attempts of Starship launches only partially successful) while all other American space companies stopped at 7 and the Europeans had to settle for 3. SpaceX, alone, did better than the sum of China and Russia, respectively with 67 and 19 launches. 2024 saw a very similar situation in the global panorama but with a further increase in the number of SpaceX launches that reached a total of 138 (132 Falcon 9, 2 Falcon Heavy and 4 Starship tests) accounting for a staggering 95% of the grand total of 145 American launches. Apart from being a customer of itself for the numerous launches of Starlink satellites, SpaceX serves a wide range of clients ranging from NASA, to the Pentagon, to ESA, to EU Commission, to large telecommunications companies, to competitors for orbital internet like OneWeb and Jeff Bezos's Kuiper. Both were forced to turn to SpaceX, owing either to the lack of Russian launchers (out of the market due to the Ukrainian crisis), or to the delays of alternative launchers like Blue Origin's New Glenn and the European Ariane 6.

In addition to populating the satellite network, it is necessary to replace the satellites that break down and reach the end of their orbital life, which is estimated to be around

3–4 years. Starlink satellites represent well more than half of the active orbital population and, being all at the same height, they are literally filling the available orbital volume so much that their automatic control system must make many anti-collision maneuvers, to avert the danger of orbital collisions.

As in the case of cubesats, we are talking about a very recent revolution that began in 2019 with the first launch of a group of 60 first-generation Starlink, strange mini-satellites shaped like tables about 3 m long with a solar panel three times as long. Not exactly small in size, the tables contain antennas that communicate with ground terminals to provide a download speed of 200 Megabits per second. The service began in the United States in 2020 and is now available in 118 nations. By the end of 2023, the satellite internet service had 2.2 million subscribers, less than initially predicted by SpaceX in the development plan, but the numbers are growing: by the end of 2024, subscribers had reached 4.6 million. Each of them, after purchasing a $499 antenna for connection with the satellites, pays around $75 a month; obviously, military, telecommunications companies, airlines, shipping companies, businesses, cruise ships, and government offices pay significantly more expensive subscriptions.

In the most remote areas of the world, Starlink offers the only connection possibility and, when the service starts to operate, it changes lives. This was experienced by some villages in the depths of the Amazon forest where, thanks to a benefactor, the antenna to receive the signal (complete with solar panel) arrived. While being connected to the rest of the world has brought undeniable benefits, it was necessary to limit the connection time to a few hours a day to prevent all the kids from deserting normal activities being glued to their phones.

Space Economy and Geopolitics

No event has demonstrated the power of Starlink—and the influence of Elon Musk—more than the war in Ukraine. Over 42,000 Starlink terminals are now used in Ukraine by the military, hospitals, businesses, and humanitarian organizations. During the heavy Russian bombings at the beginning of the invasion, which caused widespread blackouts, the Ukrainian government turned to Musk to stay online. Starlink entered Ukraine in February 2022, when Russia began the invasion and a cyber attack—later attributed to Russia—knocked out a satellite system operated by high-speed communications company Viasat that was used by the Ukrainian military. With troops and commanders offline, the Minister for Digitalization Fedorov via Twitter sought help from Musk, who, within a few hours, announced that Starlink had been activated in Ukraine and that the first shipment of terminals was in it way.

That message made Elon Musk realize that the crisis was an extraordinary opportunity to demonstrate the strategic value of his Starlink's orbital internet service, which can ensure network connection even when ground structures are damaged or destroyed. By giving away or, as it was discovered later, selling to third parties (who then donated to Ukraine) thousands of Starlink terminals, Musk got great publicity while providing Ukraine's connection to the rest of the world. Access to the network allowed the downloading of data from satellites observing the territory and enabling the tracking of troop movements to anticipate the opponent's moves and plan military actions, perhaps through drones.

The technology, which can be used anywhere, in forests, fields, villages, and mounted on the roofs of military vehicles, gave the Ukrainian army a great advantage over the

Russian forces because it allowed artillery teams, commanders, and pilots to simultaneously watch drone footage while deciding online how to move and what to hit. Response times, from the search for a target to the actual attack, were reduced from 20 min to about 1 min, and the results were clear and tangible.

At a certain point, however, Musk began to express concern about the offensive use Ukrainians were making of satellite data which, according to him, had been granted to them for defense and not for attack. His position was explained on Twitter where he wrote that the terminals "like other commercial products are for private use and not military, but we have not exercised our right to turn them off. We are trying to do the right thing, where the 'right thing' is a very difficult moral question".

The position of SpaceX is well described in some statements by Gwynne Shotwell, President of the company, who, at the beginning of 2023, said she was happy to be able to help the Ukrainians in their fight for freedom, but emphasized that Starlink should not become a weapon. Using Starlink for communication between military personnel is fine, but the service should not be used for offensive purposes, for example to guide drones that strike the enemy. Without giving details, Shotwell stated "there are things we can do to limit this use, and we have done it". In the areas where fighting is taking place, for example, *geofencing* (a geographical barrier) is applied to delimit the territory from which the Ukrainians can use the receivers. In the case of offensive actions that penetrate into the territory occupied by the Russians, it is necessary to ask Starlink to modify the geography of the connection, otherwise the advancing units have no way of communicating with each other. Geofencing is designed to prevent Ukrainians from using Starlink to guide drones into Russian territory and certainly does not give them coverage over Crimea. To counter the competitive advantage offered by the availability of the Starlink connection, the Russians have learned to disrupt the signals but do not disdain to also resort to Starlink terminals purchased on the black market.

The story of Starlink's involvement in Gaza is different, but here too the weight of Musk in the global geopolitical framework emerges. On October 28, 2023, while the Israeli army's offensive on Gaza was underway, US House Democrat Alexandria Ocasio-Cortez tweeted "Cutting off all communications with a population of 2.2 million people is unacceptable. Journalists, medical professionals, humanitarian workers and innocents are all in danger". She then continued "I don't know how one can defend such an act. The United States has historically denounced this practice". Perhaps she was struck by the many messages from people who could no longer contact their relatives in the Gaza Strip, since communications had been interrupted following air and ground attacks in the region.

Even humanitarian organizations were no longer able to communicate with the teams present in the area. The World Health Organization had declared that it was not "in contact" with its staff in Gaza and the Commissioner-General of the United Nations Relief and Works Agency for Palestine Refugees had written to the staff stating that the organization was "deeply concerned" for them also owing to the blackout situation.

Surely, the congresswoman was not addressing Musk directly, but he responded to the post, saying: "Starlink will support connectivity to internationally recognized humanitarian organizations in Gaza".

Perhaps it was meant to be an act of goodwill, or perhaps it was the expression of the visionary entrepreneur who envisioned satellite internet connection and who now feels invested with the role of Deus ex machina in war zones since, with the magic wand called Starlink, he can give (or take away) the internet connection.

No one from Gaza had asked him anything (unlike what had happened for Ukraine) and he himself shortly after declared that there was no Starlink traffic in the Strip. However, his willingness to provide connection to international organizations triggered the reaction of Israel government, which promised to do everything to prevent connectivity from being restored because it feared that Hamas could use it. The Israeli Minister of Communications, Shlomo Karhi, had stated that he intended to cut all ties with Starlink. The problem was sorted out at the end of November with the visit of Elon Musk to Israel, whose government agreed that Starlink could provide connections to authorized operators.

Space Monopoly

When a single man can decide to deny access to Starlink for a customer or a country and, if he wanted, could have the ability to exploit sensitive information collected by the service, it is time to ask some questions. It should be emphasized that Musk has done nothing illegal, he has simply seized an opportunity in the field of global internet connection by exploiting the lack of rules in the space economy sector. His company, SpaceX, whose value is estimated at around 350 billion dollars (almost twice the evaluation in 2023), is able to carry out the vast majority of American launches. Moreover, his Starlink satellites are already well over half of the active satellites and, for sure, represent the vast majority of low orbit satellites.

Moving as a master in the space field, and keeping rhythms that no one can replicate, Musk is effectively a monopolist, in the field of launches and satellite internet connection services as the war in Ukraine has tragically shown.

I do not believe that, when he designed Starlink, Musk thought of assuming such an important role in the global geopolitical framework. I think he saw in this great Space Economy project interesting prospects for profit, then came the international crises and the orbital connection took on an extraordinary strategic value. In all respects, Musk is an anomaly in the global geopolitical framework that has never seen so much power in the hands of an individual who has recently acquired an important political role in the new Trump administration and who, with a message, can create (or mitigate) international crises.

And the Others?

Even though, at the moment, Musk is the only entrepreneur able to offer global internet connection, the situation will evolve. Other companies also plan to offer internet from satellite, but their competition is still far away. The multinational OneWeb might have another marketing strategy and does not seem to aim at private users, but rather at television networks or service providers, while the Kuiper constellation of Jeff Bezos's Blue Origin has only launched two test satellites, whose launch was entrusted to SpaceX because the Blue Origin launcher was late.

However, more competing companies mean many more satellites: based on the authorization of the USA Federal Communications Commission (FCC) SpaceX has the green light to launch a constellation of 12,000 satellites, OneWeb has a constellation of 650 satellites, then we must consider the 3200 planned for the Kuiper project, the 4700 of Samsung, the almost 3000 of Boeing. Russia and China certainly do not want to miss the opportunity and they also plan to have their constellations to provide internet from orbit to the inhabitants of their vast territories. As a matter of fact, on Aug. 6th 2024, China has launched the first group of 18 satellites belonging to the Thousand Sails constellation which envisions 15,000 satellites. Not to mention that Elon Musk has always said he wants to expand Starlink to reach 42,000 satellites and has already asked for, and obtained, from the FCC authorization for the launch of a first batch of another 30,000 satellites. Adding up, even considering that perhaps not all projects will be realized, we easily reach 100,000 new satellites in low orbit just for the Internet service.

And this is where the problem arises.

The circumterrestrial space, as large as it is, is not infinite and may not be able to ensure vital space for all satellites.

The Environmental Impact of the Growth of the Space Economy

The undeniable success of space services also has some negative aspects that we should worry about before the situation gets out of control.

The ecological footprint of a space mission includes its entire cycle encompassing design, construction, launch, in-orbit management, and end of orbital life.

Neglecting, for the moment, the impact of the satellite's construction, let's consider what happens during the launch, the orbital life, and the reentry of the satellite.

The Launch and the Atmosphere

The impact of a launch begins with the disturbance it causes to the management of airspace. Since it is certainly not advisable to risk close encounters between flying aircraft and ascending rockets, each launch implies the temporary closure of the affected airspace. The explosion of the StarShip capsule on Jan. 16 2025, shortly after the launch from Boca

© The Author(s), under exclusive license to Springer Nature Switzerland AG 2025
P. Caraveo, *Space Ecology*, https://doi.org/10.1007/978-3-031-78344-9_4

Chica, in Texas, forced air trafic control to close the airspace above the Caribbean islands where debris were falling. For launching sites near the ocean, it is also necessary to control maritime traffic because, in case of problems, material may fall into the water. Maritime control is a very sensitive issue for the Tanegashima base in Japan. Being a peninsula in a location heavily frequented by commercial fishing boats, the Japanese space agency (JAXA) must negotiate with the fishermen's union the launch windows to minimize the interruption of activities.

In addition to these managerial/economic impacts, there may be environmental impacts, especially during test flights of new launchers. This happened when the first test of SpaceX's Starship launcher was attempted from the Boca Chica base in Texas. Having pressing deadlines with NASA's lunar program, Elon Musk was in a terrible hurry to carry out the first test flight and decided to do so even though the system to protect the launch pad from the terrible heat produced by the 33 engines of the heavy launcher at takeoff had not yet been installed. The test on April 20, 2023, which failed after a few minutes, destroyed the concrete base, raising a huge dust cloud that also reached the inhabited centers of the area, which did not appreciate it at all. Moreover, Boca Chica, like Cape Canaveral, is located within a wildlife oasis and, to obtain the authorization to carry out a new test, SpaceX had to demonstrate to the U.S. National Park Service that the steel protection plate, cooled with a waterfall, would prevent the recurrence of accidents of similar proportions. In fact, since the tests of the various components of the Starship vector began in 2019, at least 19 accidents have been recorded that have caused fires, explosions, fuel leaks that certainly were not beneficial for the wildlife oasis. The Fish and Wildlife Service, which operates within the National Park Service, had to accept the fact that the Federal Aviation Administration, which must ensure the safety of launches, does not consider environmental protection its most important priority. They are certainly not happy if a rain of debris falls into national park, but until there are accidents with deaths and injuries, they close both eyes in the name of the progress of SpaceX's projects, which are considered really important for the American space program.

Let's now come to the actual launch, which burns tons of fuel releasing potentially polluting gases into the atmosphere. When a rocket like the Falcon 9 takes off, it generally takes about 90 s to cross the lower atmosphere, or troposphere, before reaching the middle atmosphere which is, in general, an uncontaminated place, not reached by terrestrial pollution. It is therefore calm, except for the occasional rocket, which crosses it for 3 or 4 min on its journey to space. By the time a rocket enters orbit, it will have discharged into the middle and upper layers of the atmosphere about two-thirds of its exhaust gases, which will later gather in the lower layer of the middle atmosphere, the stratosphere, which hosts the ozone layer that protects us from the harmful radiation of the sun.

Even though today the exhaust gases of rockets pale in comparison to those emitted by aviation (which, however, discharges them at much lower altitudes), there is a well-founded fear that small additions to the stratosphere could have a very important effect since the ozone layer is extremely sensitive, and even a minimal change can have enormous effects. Unfortunatey, the Montreal Protocol, which successfully set limits on chemicals known to damage the ozone layer, does not address the emissions from rockets or satellites because in 1987, when the document was approved, the number of launches was much smaller than today and this problem did not arise.

Just like our cars, exhaust gases depend on the fuel used, but we know they contain nitrogen and chlorine oxides capable of reacting with ozone, destroying it and thinning the layer of this gas so important for protecting us from the Sun's ultraviolet radiation. For its Falcon 9, SpaceX uses a fuel similar to kerosene that deposits large amounts of soot in the stratosphere, 500 times more efficient in absorbing solar radiation compared to emission of planes that fly much closer to the surface. With the number of launches continually increasing, there is concern that the soot, released by current rockets burning hydrocarbons, could raise the temperature of the troposphere by 2° with serious consequences for the ozone. Even "greener" launchers that burn liquid hydrogen would have a negative effect because the water vapor produced by combustion is a greenhouse gas, especially in such a dry environment. Unfortunately, what is "green" near the ground may not be so in the upper atmosphere, which, with its ozone layer, is so important for life on Earth. While the impact of a single launch could be considered irrelevant, what really matters is the number of launches performed at global level. Figure 1 shows the impressive growth we are witnessing and the clear prominence of the american contribution.

To get an idea of the magnitude of the problem, consider that we are talking about having 100,000 satellites in orbit with an orbital life that can range from 5 to 10 years. This means that every year roughly 10,000 satellites will have to be launched to replace those that have reached the end of their life and will have to be deorbited. Assuming that each launch brings 50 satellites into orbit, at least 200 launches per year will have to be carried out just to maintain the satellite internet network. Other launches will then be necessary to meet all other civil and military service requests such as Earth observations, weather forecasts, border surveillance. We are talking about hundreds of launches that will burn large amounts of fuel spreading exhaust gases that will heat the stratosphere accelerating the chemical reactions that destroy the ozone. Unfortunately, it is possible that some of these effects are already present in the northern hemisphere, from where most of the launchers depart. During the winter and spring of 2020, the largest and

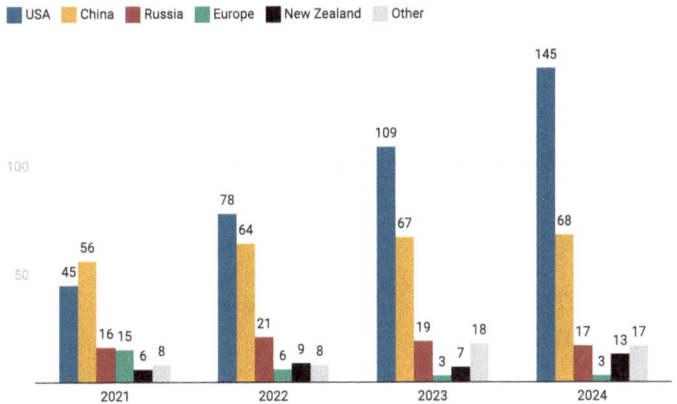

2024 Orbital Launch Attempts by Country

263 orbital launches were attempted last year. 258 reached orbit/near orbit.

■ USA ■ China ■ Russia ■ Europe ■ New Zealand ▢ Other

The data was compiled by astronomer Jonathan McDowell. Note Rocket Lab missions in New Zealand are not considered US launch. Starship launches are considered orbit/near orbit. The 2024 other category includes Japan (7), India (5), Iran (4), and North Korea (1).

Created with Datawrapper

Fig. 1 Number of launches in recent years divided by the nationality that carried them out. (Credit Payloadspace.com). The growth in the number of launches is impressing and every year beats the record of the previous one

longest-lasting ozone hole ever recorded was measured over the Arctic, which Earth's atmosphere experts could not explain. Could it be due to SpaceX launches? If true, it would be time to start worrying.

Certainly, the environmental impact of launches falls into a gray area where there are no rules.

As described in the case of environmental damage related to the activity of the Starbase in Boca Chica, in the United States, the Federal Aviation Administration evaluates the environmental impacts of rocket launches on the ground, but not in the atmosphere or in space. The Environmental Protection Agency is not responsible for analyzing rocket launches. The Federal Communications Commission grants licenses to large satellite constellations but does not consider their potential environmental damage, because the

FCC is exempt by law from making environmental impact assessments. It is clear to everyone that the increase in the number of launches can only worsen the problem of gas and soot pollution of the upper atmosphere, and it would be unwise to wait to take action when the damage will be too difficult to remedy. It would be much better to accurately assess the problem now to try to limit the environmental consequences of the new space race.

The Atmosphere Does Have an Impact on Launches

The Earth's atmosphere is not always the same: its density, which varies depending on the altitude, can be modified in "response" to the activity of the Sun. Even though our star is generally quiet, it can occasionally produce spectacular surface explosions called Coronal Mass Ejections, during which a huge amount of energy is released in the form of accelerated particles that propagate in the interplanetary space and that, if they hit the Earth, can cause a *solar storm*. An event that can be predicted and that, normally, is not worrying, unless you are an astronaut or are about to launch satellites. Indeed, the injection of energy from the Sun does modify our atmosphere, which swells changing its density profile. This is a very critical parameter when you want to insert satellites into low orbit where the atmosphere, although very rarefied, is still there. SpaceX learned this at its own expense, when on February 3, 2022, it lost 40 of the 49 just-launched Starlink satellites because it had not taken into account the solar storm of the previous day. Starlink satellites are released, one after the other, at an altitude of 210 km. This is a lower quota than the operational one of 500 km and was chosen as a precautionary measure. If a satellite does not pass the first orbital tests, from the 210 km quota it is easy to de-orbit it to make it burn in the atmo-sphere without leaving space debris around. However, the procedure that works well when the Sun is quiet proved disastrous on February 3, 2022, when the atmosphere was denser than usual and friction was a real killer (Fig. 2).

Having learned the lesson, Musk paid a lot of attention to the intense magnetic storm that hit the Earth on the night between May 10 and 11, 2024. In a Tweet, where he showed the graph of the geomagnetic disturbance that had reached the maximum level of the scale, he said that the satellites in orbit were resisting the pressure. You don't mess with the Sun. But it's not just solar activity that can cause the unexpected re-entry of Starlink satellites. On July 11, 2024, a malfunction of the second stage engine caused the release of 20 satellites at too low an altitude. Even though SpaceX tried to activate the satellites' ion engines at their maximum power, atmospheric friction got the better of them and the 20 satellites sadly re-entered and were destroyed.

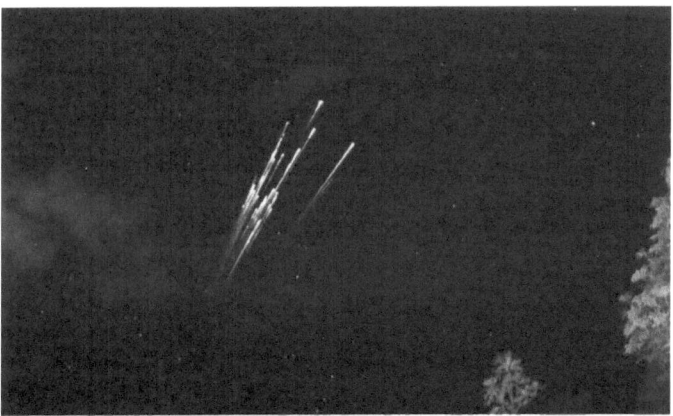

Fig. 2 Flaming re-entry of the Starlinks that failed to go into orbit on February 3, 2022. (Credit Caribbean Astronomical Society)

In-Orbit Management

Space is becoming increasingly crowded and the rapid growth in the number of satellites has immediate implications for the sustainability of the use of Earth orbits. Even though at first glance it might seem that space is so large that there could not be overcrowding problems, unfortunately this is not the case. The most interesting orbits from a commercial point of view are the geostationary ones, where the large communication satellites operate, and those closest to the Earth, let's say between 300 and 600 km in height, where Earth observation satellites (civil and military), various cubesats and large constellations for global connection crowd together. While, as we have seen, the use of the geostationary orbit is regulated by precise rules, for objects in low orbit there are no international rules and growth is so rapid that an orbital traffic jam situation is feared, with increasingly frequent approaches between satellites that risk collisions with each other as well as with all the scrap metal, large and small, that we have sent into orbit over the past 60 years. The peak occupation of the orbital shell at 500 km high, due mostly to Starlink satellites, is clearly a source of concern since, according to various studies, it is already occupied for about half of its maximum capacity and the risk of saturation cannot be neglected (Fig. 3).

While part of the world praises Elon Musk for the visionary idea of global internet, the 18th squadron of the Space Control within the U.S. Space Force sees it as a lot of extra work, since they are the ones who have to oversee space traffic keeping under control more than 30,000 objects larger than 10 cm. For each one the orbit and the expected position is calculated, moment by moment, and compared with that of all the others. When it is expected that the distance between two objects will be less than 1 km, the operators receive an alert message because they must pay attention to

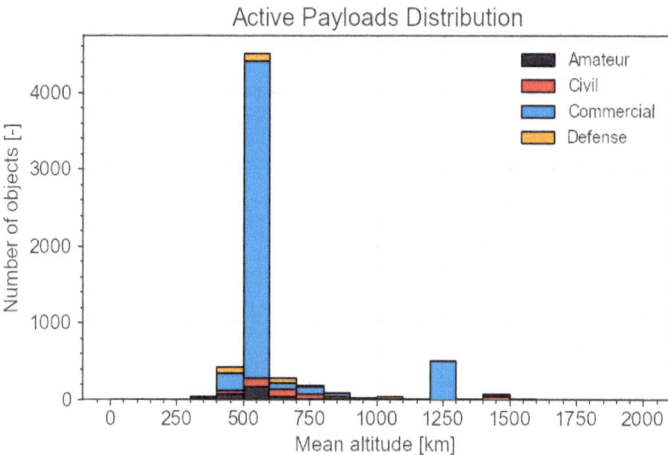

Fig. 3 Number of operational LEO satellites in different orbital shells. The peak at 500 km is due almost entirely to Starlink satellites. (Source ESA Space Environment report, 2023)

the evolution of the situation. If it seems excessive to worry about an approach of about 1 km, consider that the low orbit satellites move at 8 km per second and that their position is known with a precision of 100 m. Any error could be fatal.

Precisely for this reason, the rapid increase in the population of satellites in low orbit is likely to create quite a few problems, also because everyone is aware that this is just the beginning. The numbers will certainly grow. SpaceX has been authorized to launch 12,000 satellites, a number greater than everything that had gone into orbit from the beginning of the space age to the time of such authorization.

A future with heavily trafficked orbits is foreseen, which, among other things, amplifies the possibility of collisions between satellites with chain effects that are quite worrying. In fact, in case of collision, a cloud of debris is formed that will continue to follow the same orbit of the original satellite multiplying the probability of other impacts. It is called Kessler syndrome from the scientist who first studied it. Despite the apparent vastness of circumterrestrial space, we know that the probability of collision is not zero and, on February 10, 2009, we observed a textbook one, which will be described later.

Thus, a constant surveillance to minimize the risk of dangerous approaches is badly needed. The growth of the number of satellites in low orbit represents also a serious problem for astronomers. Satellites' metallic structures reflect solar radiation making them bright sources in the night sky, responsible for bright "streaks" in astronomical images, which risk being irreparably dirtied by this new type of light pollution that spares no place on our planet. Satellites are also a source of considerable radio noise which is far brighter than the radio emission of celestial objects, thus hampering radio telescopes' observations. In addition to disturbing astronomical research, the proliferation of the population of satellites in low orbit increases the number of close encounters which currently amount to hundreds per day, but their number is constantly growing.

Between December 1, 2022 and May 21, 2023, Starlink satellites were forced to move more than 25,000 times. It's about twice the moves in the previous semester, from June to November 2022. Overall, the SpaceX satellites had to perform these automatic maneuvers 50,000 times (Fig. 4).

The fear, however, is that the number will continue to rise dramatically with SpaceX planning the deployment of its Starlink 2.0 satellites and with other companies like Amazon wanting to launch the Kuiper constellation, the Chinese Thousand Sails, and the already operational OneWeb.

With the number of anti-collision maneuvers seeming to double every 6 months, if current trends continue, Starlink

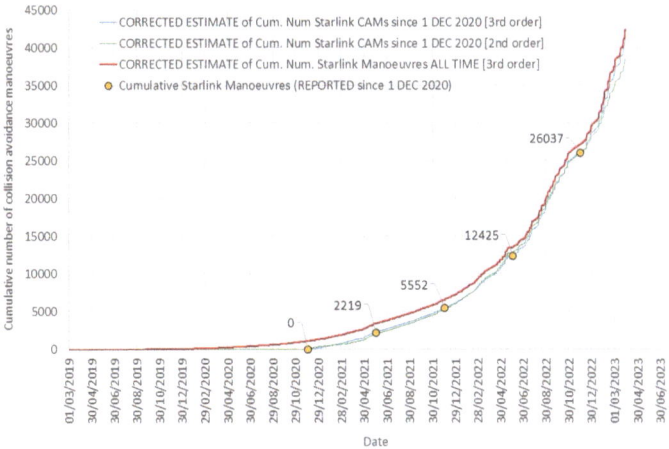

Fig. 4 Cumulative number of anti-collision maneuvers performed by Starlink satellites. Their number doubles every 6 months. (Credit Hugh Lewis https://www.sankyungroday.com)

satellites will have to perform about a million maneuvers every 6 months by 2028.

Most are approaches between two Starlinks, but this does not worry SpaceX because the satellites have an anti-collision system and can move autonomously to get out of the way if they detect something nearby. This is certainly positive for Starlink but does not excite space traffic controllers who have to manage satellites that change orbital parameters, thus requiring an update of all projections of future positions.

About 500 alerts each week involve a Starlink and a different object. If it happens to be an active satellite, its managers have to devote time and attention to making (and remaking) projections on the objects' relative positions before deciding whether to move the satellite, or whether it is appropriate to contact SpaceX. In the end, it's a risk calculation problem. Certainly, one wants to avoid the danger of collision that would destroy the satellites creating a huge number of extremely dangerous debris, but, on the other hand, one cannot continuously maneuver the satellites, a procedure that requires time, planning, fuel, and temporarily interrupts services.

Space Junk

Earth orbits are also crisscrossed (and occupied) by a very high number of debris known as space junk. The American Department of Defense's Space Surveillance Network (SSN) tracks tens of thousands objects larger than 10 cm and provides alerts in case of dangerous approaches.

Space junk encompasses non-functioning satellites, stages of carrier rockets that have pushed their load into orbit but then remained themselves to orbit, explosive bolts, tools lost by astronauts engaged in ISS maintenance work, pieces of satellites exploded either accidentally or otherwise.

The veteran of defunct satellites is Vanguard 1 which went into orbit, along with the last stage of its launcher, in 1958. Considering that it describes an ellipse with an apogee of 3969 km and a perigee of 650 km, it will continue to orbit for at least a century.

According to the census as of February 2nd 2025, by J. McDowell's site (https://planet4589.org/space/stats/active.html), in addition to the 11,108 operational satellites there are more than 3000 defunct satellites, almost 2000 launchers, 1500 pieces of satellites (presumably exploded) and more than 12,000 cataloged debris coming, for the vast majority, from a Chinese anti-satellite test and the collision between two satellites.

Data as of	2025 Feb. 2
Active Starlinks, orbit:	6935
Other active maneuverable payloads, orbit:	2423
Active non-maneuverable payloads, orbit:	1750
Total all active payloads, orbit:	11108
Dead Starlinks, orbit:	34
Other dead satellite payloads, orbit:	3036
Dead payloads, orbit, marginal cases:	21
Total all active and dead payloads, orbit:	14199
Rocket stages, orbit:	2006
Sat components, orbit:	1474
Cataloged debris, orbit:	11597
Total cataloged objects, orbit:	29636

Considering the masses of all objects in orbit, active and not, we find that the mass of active instruments is about half of the total. In other words, about 6000 tons of junk orbit in circumterrestrial space.

Data as of	2025 Feb. 2
Tonnage of active Starlinks in orbit:	3316.7
Tonnage of other active maneuverable payloads in orbit:	4040.0
Tonnage of active non-maneuverable payloads in orbit:	354.7
Tonnage of all active payloads in orbit:	7711.3
Tonnage of dead Starlinks in orbit:	25.9
Tonnage of other dead satellite payloads in orbit:	2188.7
Tonnage of dead payloads in orbit, marginal cases:	5.6
Tonnage of all active and dead payloads in orbit:	9931.5
Tonnage of rocket stages in orbit:	3809.6
Tonnage of sat components in orbit:	162.2
Total cataloged tonnage in orbit:	13903.3

According to the census conducted by the European Space Agency in September 2024 https://www.esa.int/Space_Safety/Space_Debris/Space_debris_by_the_numbers, in addition to the nearly 37,000 objects that are continuously monitored (active and dead satellites, spent rockets and debris of various origin both known, thus catalogued, and unkown), the number of smaller objects, which escape measurement, could easily be over 100 million.

Continuous monitoring, in addition to recognizing each object, notices when something has happened in orbit that resulted in a sudden increase in the number of debris. Thus we know, obviously in retrospect, when military tests known as DA-ASAT (for Direct Ascent Anti-Satellite) are conducted. During such tests a missile launched from the ground destroys a target satellite, producing a cloud of debris that continues to move following the orbit of the satellite from which they originate. Over the years, these demonstrative actions have been conducted by China (in 2007), the USA (in 2008), India (in 2019) and Russia in November 2021, as clearly seen from the graph showing the number of tracked orbiting objects over time (Fig. 5). The most evident growth is due to the 2007 Chinese test, which, by destroying a satellite at 700 km altitude, produced 2000 fragments, most of which are still dangerously in orbit. Conversely, the anti-satellite tests carried out by the USA and India destroyed objects at less than 300 km altitude and the debris were cleaned up by the atmosphere.

The story of the DA-ASAT of November 2021, which occurred at an altitude compatible with that of the Space Station and generated a serious state of alert on board, is different. While the U.S. Space Command was trying to track the orbit of the 1500 debris created, the occupants of the ISS (astronauts and cosmonauts) were awakened and had to put on their flight suits and board the Crew-Dragon and Soyuz spacrafts, which had to be activated to be ready to detach in case of impact. After a few hours the alert was called off but the incident created quite a friction between Russians and Americans. In the end it was understood that even Roscosmos, the Russian space agency, had not been informed of the test decided by the Ministry of Defense without taking into consideration the safety of the Russian cosmonauts. Not to mention the other occupants of the ISS

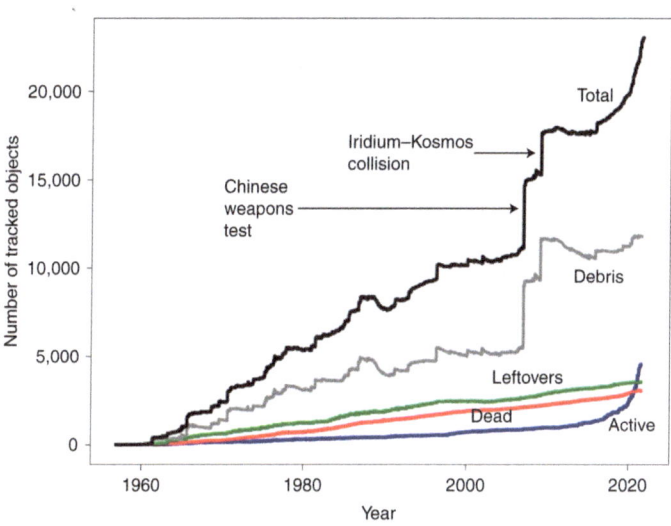

Fig. 5 Number of monitored objects over time: in blue the active satellites (with the rise of recent years), in red the dead ones, in green various kinds of leftovers, mostly pieces of launchers, in gray the space debris and in black the total curve. The debris curve shows how their number is decidedly predominant compared to all other components of the orbiting object population. (From A. Lawrence et al "The Case for Space Environmentalis", licensed under CC BY 4.0)

and the Chinese taikonauts on board the Tiangong station. DA-ASATs are not offensive actions in a military sense because the satellites destroyed are debris produced by the same nation conducting the test, but their effect is devastating for the orbital ecosystem. Consider that a few days before the Russian test, the ISS had to raise its orbit by a couple of kilometers to minimize the risk of collision with one of the debris created by the Chinese test of 2007 which, after over 15 years, continue to orbit and pose a potential danger (Fig. 6).

Obviously, the number of debris can suddenly increase due to the accidental explosion of a satellite as happened on June 26th, 2024 when the Resurs P1 satellite, launched in 2013 by Russia to observe the Earth and active until 2022,

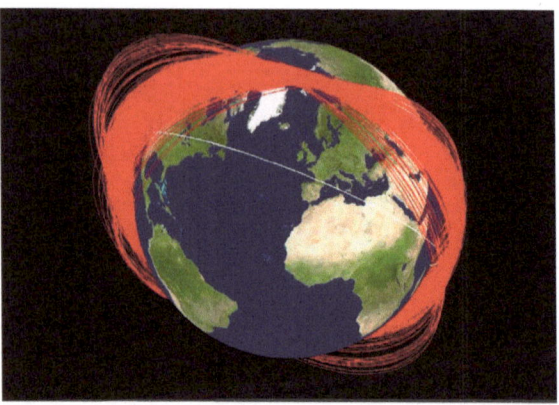

Fig. 6 The over 2000 traceable debris of the Fenyun-1 satellite 1 month after its destruction in 2007. In white the orbit of the ISS. (NASA Orbital Debris Program Office)

exploded breaking into at least 100 fragments that were deemed potentially dangerous for the nine people aboard the ISS who had to implement, as a precaution, the emergency procedure and take their places on the three spacecrafts (Soyuz, Crew Dragon, and Starliner) attached to the ISS to be ready to detach in case of impact. The alert lasted about an hour, but the explosion of June 26th is not an isolated phenomenon. About 10 are counted each year and, in general, it is the fault of the batteries or residual fuel, as was probably the case for the destruction of the second stage of Long March 6A which exploded on Aug.6th after the successful release of the first 18 satellites of the Chinese constellation Thousand Sails. But it can also be an unfortunate orbital coincidence when two satellites orbit too close to each other as happened at 16:56 UTC on February 10, 2009 at about 790 km altitude above Siberia, when Iridium 33, a telecommunications satellite operating within the IRIDIUM constellation for global telephony, found COSMOS 2251, a Russian satellite launched in 1993 and inactive for about 10 years, in its path (Fig. 7).

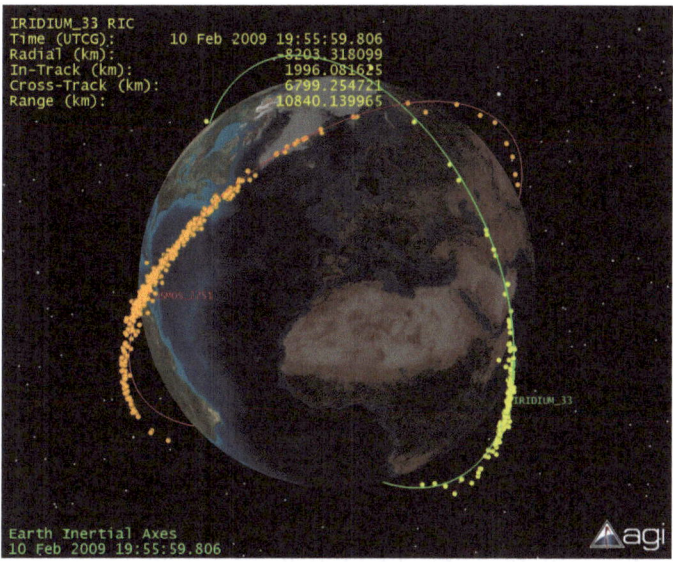

Fig. 7 Three hours (thus, two orbits) after the collision this was the distribution of the debris of Iridium (yellow) and Kosmos (in orange). (Credits: Celestrak/AGI Viewer 9)

Both satellites were tracked by the radars of the USA strategic command center. Based on the mapping of their orbits (both of polar type, roughly perpendicular to each other) it was known that they would approach but if a dangerous situation was feared, the Iridium 33 satellite could have activated its engines to get out of the way. At the moment of the crossing, however, they collided. While the managers of the IRIDIUM system were losing the satellite signal, those at the monitoring center saw hundreds of dots appear on their radars. Immediately 355 fragments of COSMOS 2251 and 159 of Iridium were counted. In addition to being more numerous, the fragments of COSMOS are more scattered in height going from 198 to 1689 km. The fragments of Iridium are confined between 582 and 1262 km. While the greater mass of the COSMOS satellite can explain, at least in part, the more substantial production of debris, the scattering in height must be partly due to the fact that COSMOS was a pressurized satellite, and that therefore the collision caused a real explosion. The first fragments of COSMOS began to re-enter the atmosphere a month after the collision. While the fragments that burn in the atmosphere are not a cause for concern, those still in orbit, in the range of heights they occupy, can pose a potential danger to other missions. Following the incident, in 2009 the probability of collisions at the height of the space station increased to 1 chance in 300. A value still lower than NASA's safety limit which is 1 in 200, but not so far away. Fortunately, the atmosphere provides a natural cleaning mechanism for low orbits.

What Can Be Done to Prevent This from Happening Again? First of all, it would be desirable to have only active satellites in orbit. Those that have completed their missions should disappear without causing damage. This can be achieved with a dedicated system (engine plus fuel) to make a controlled re-entry into the atmosphere. Obviously, this is a solution (which unfortunately has costs) to be adopted for new satellites. The thousands of objects launched since the beginning of the space age to date, for the most part, remain and will remain where they are. Only those on relatively low orbits will gradually be slowed down by friction with the high atmosphere and will re-enter.

Clearly, it is imperative to avoid adding space junk, whether they are debris resulting from accidental or voluntary explosions, pieces of the last stages of satellites, equipment lost by astronauts. Every useless object in orbit is a potential danger to astronauts and other satellites.

The decision of agencies to self-impose the reentry of satellites at the end of their orbital life is an important step in the right direction. Unfortunately, not much can be done for what is already in orbit: there are many projects for the collection and disposal of orbital debris, and several missions are planned to test alternative methods for capturing and de-orbiting cosmic wreckage. To capture a passive and uncontrollable wreck, first of all, it is necessary to be able to approach it safely. This is what the ADRA-J satellite of the private Japanese company Astroscale did on May 23, 2024, when it came within 50 meters of the exhausted upper stage of a discarded Japanese rocket which launched an Earth observation satellite in 2009 (Fig. 8).

The photographic inspection allowed to verify the conditions of the object and its structural integrity, valuable information to decide how to proceed with the capture of the wreck. This will be the purpose of a new mission by Astroscale that will use a robotic arm similar to the one used on the International Space Station, but lighter. It is therefore clear that the problem of debris removal is not technological, but rather economic/political. The missions are expensive and it is not at all clear who should bear the costs of

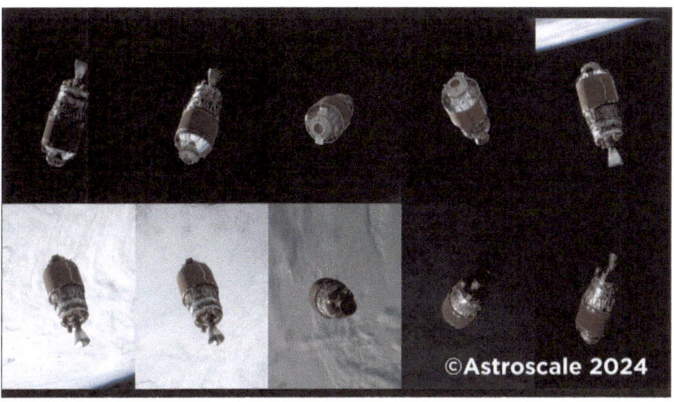

Fig. 8 Close-up view of a space debris. (Credit Astroscale)

orbital cleaning. All international treaties that regulate space disputes (reported in the appendix) do not cover these issues, which are instead the subject of rules that the major space agencies have voluntarily accepted in recent years. Almost all of the objects that have no chance of de-orbiting were launched before the agencies decided on the anti-debris rules, so the state or company that launched them has not violated any rules and therefore cannot be forced to pay anything.

Re-entry into the Atmosphere

As we have seen in the description of low orbits, it is the friction, very light but continuous, due to the very thin residual atmosphere, that is responsible for lowering the orbit and, in the end, for the destruction of satellites. The re-entry can be controlled, if the satellite (or the piece of launcher) has a propulsion system and enough fuel for the maneuver, or uncontrolled, when the object is passive. As shown in Fig. 9, the first to detach are the solar panels that offers the

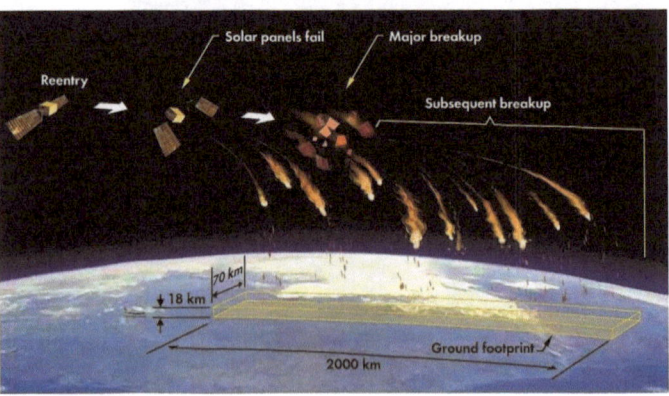

Fig. 9 Diagram of the fragmentation of a satellite destroyed by friction with the atmosphere. (Credit Australian Space Academy)

most surface to friction, while the body of the satellite breaks up and burns later. The material that re-enters in an uncontrolled way amounts to about 2 tons every 2 weeks. Unless it is a controlled re-entry, designed to occur over a particular point in the Pacific Ocean known as point Nemo, chosen so as not to cause damage in case some part of the satellite survives the burning entry into the atmosphere and reaches the ground, it is never possible to say precisely when a satellite will re-enter or over which part of the globe it will happen. The area where the debris from the fragmentation could potentially fall is a few hundred km long and 70 km wide. Fortunately, most of our planet is covered with water, but some space scrap, typically tanks and batteries, can pay a visit as happened on March 8, 2024 to Alejandro Otero of Naples, in Florida, who had his house damaged by space debris. Mr. Otero posted on social media the photo of the artifact asking for help to get in touch with NASA because, from what he could read, he suspected they knew something about it.

It didn't take long to understand that it was a piece of a lithium battery weighing just under a kilogram, part of a much larger battery that provided power to the Japanese module located outside the space station. When the platform was decommissioned, detached from the ISS and left to its fate in March 2021, everyone thought it would be destroyed during re-entry into the atmosphere.

Instead, something survived and now Mr. Otero is wondering who will pay for the damage. At first glance, it would seem a case covered by the Convention on the responsibility for damage caused by space objects of 1972. As described in the Appendix, the responsibility falls on the country that launched the object, regardless of whether the battery was part of the Japanese module. The launch took place from the USA so it will be US's government to compensate. While every now and then there are concerns for the uncontrolled re-entries of pieces of launchers, or large satellites, which may start to rotate in a disordered way making even more difficult to predict the area of a possible impact, it is curious that the only confirmed damage from a space debris was caused by an harmless object which was not considered dangerous.

Although this is the only example of damage to private property caused by space debris, the piece of battery is certainly not an isolated case. Much larger and heavier debris rain from the sky around the world. In some cases, these are examples of unexpected survival of pieces that were supposed to burn in the atmosphere, in others, however, the arrival of the space wreckage is an unpleasant certainty.

Let's start with surprise re-entries like those due to the "trunk" of the Dragon capsules from SpaceX, both in the cargo and the crew version. This is the service module of the capsule with aerodynamic control fins, capable of transporting loads or small satellites, it is equipped with solar panels that provide energy to the spacecraft when it is in flight or docked to the International Space Station. When the spacecraft re-enters, whether it is carrying cargo or transporting astronauts, it detaches the trunk which can remain in orbit for months before making an uncontrolled re-entry during which it was expected that the structure would disintegrate. Perhaps it's the fault of the composite materials used, but it has already happened several times that portions of the trunk have reached the earth. In August 2022 a piece was recovered in Australia. Then it happened again in February and May 2024 in Canada and North Carolina where on May 22, along a pedestrian path, a piece as large as a car hood covered with a carbon fiber weave was found. In the following weeks, several smaller pieces were also found in the yards of local cottages.

Only on June 20, 2024, NASA stated that the debris actually came from the Crew Dragon 7 spacecraft confirming that this is not an isolated case. A piece of debris from the Dragon cargo mission number 30, detached from the ISS on April 28 and then splashed down on April 30, ended up in Arabia.

Clearly if a piece of debris the size of the one in North Carolina fell in a densely populated area, or on a plane, the situation would be serious and NASA, along with SpaceX, is considering options to change the re-entry procedure by keeping the trunk attached to the spacecraft for a longer time until lower altitudes. This would involve the consumption of a larger amount of propellant and it is not clear if the procedure would not impact the safety of the astronauts. For the moment being, SpaceX, in agreement with NASA, has decided that, starting from 2025, the splash down will move from the Gulf of Mexico to the Pacific coast, out of California. Changing the landing trajectory would increase the probability that trunk's debris, if any, will fall into the Ocean.

If the cases described above involve unexpected re-entries, there are many other examples of re-entries that were not a surprise because they involve pieces of launchers whose mass is such that the arrival of debris, even of large size, is certain. The launch sequences vary depending on the rocket models but almost all have a first stage that often reaches orbit where it ends up being abandoned to its fate unless it has a system to allow a controlled reentry. The difference is very important because uncontrolled reentry can occur at any point in its flight trajectory, while controlled reentry, which requires an engine and fuel, can be directed towards a remote area of the ocean.

Most rockets also have one or more "upper stages", which carry the "payload" (i.e., one or more satellites) into orbit. Of course, also these portions of the rocket must reenter, and this can happen in a controlled or uncontrolled manner.

In 2020, over 60% of launches to low Earth orbit left their rocket bodies in orbit for days, months, or even years, creating a risk of collision with operational satellites. Moreover, in the event of an explosion of the residual fuel, the rocket leftover can fragment into thousands of pieces of space debris, smaller but still potentially destructive, creating even more risks for satellites. There is also another risk due to reentry when a substantial fraction of their mass survives the heat of atmospheric friction resulting into large debris, potentially lethal that can pose serious risks on land, at sea, and for people on board aircraft.

In May 2020, the 18-tonne central stage of a Long March 5B rocket, used to launch an unmanned experimental capsule, reentered the atmosphere in an uncontrolled manner. Debris from part of the rocket, including a 12-meter long tube, hit two villages in the Ivory Coast, causing damage to several buildings. A year later, another 18-tonne central stage of a Long March 5B rocket, used to launch part of the new Chinese space station Tiangong, made an uncontrolled reentry, crashing into the Indian Ocean. These two rocket stages were the heaviest objects to reenter in an uncontrolled manner since the Soviet Union's Salyut-7 space station in 1991.

It is obvious that such reentries pose risks, but there is no international regulation that establishes what the level of "acceptable" risk should be. The United Nations Guidelines for the mitigation of space debris from 2010 recommend that the reentry of spacecraft does not pose "an excessive risk to people or things", but do not quantify this statement. The United Nations Guidelines from 2018 for the long-term sustainability of activities in outer space invite national governments to address the risks associated with the uncontrolled reentry of space objects, but do not specify how. There is no binding treaty that addresses the issue of space debris reentries, apart from the 1972 Liability Convention (see Appendix), which establishes that "the launching State is absolutely liable to compensate for damage caused by its space object on the Earth's surface or to aircraft in flight".

In the United States, the Orbital Debris Mitigation Standard Practices (ODMSPs) apply to all launches and require that the risk of accidents due to the reentry of a rocket body be below the threshold of 1 in 10,000. However, the US Air Force has decided not to apply the ODMSP standards for 37 of the 66 launches carried out on its behalf between 2011 and 2018, since it would have been too expensive to replace the rockets used (and clearly non-compliant) with more suitable ones. NASA has waived the requirements seven times between 2008 and 2018, including for an Atlas V launch in 2015 where the risk of accidents was estimated at the level of 1 in 600.

The situation is made more complicated by the fact that the risk is not the same for all inhabitants of the planet. Since many of the launchers causing uncontrolled reentries are associated with launches to geosynchronous orbits, located near the equator, the cumulative risk of reentries of launcher pieces is significantly higher in the regions close to the equator, compared to higher latitudes, where the main space powers are located.

This situation, in which the risks arising from activities in developed countries are disproportionately borne by the populations of developing countries, is certainly not new. The risk posed by the re-entry of space debris is further exacerbated by the quality of the buildings which generally offer a lower degree of protection. According to NASA, about 80% of the world's population lives "without protection or in poorly protected structures that provide limited protection against falling debris".

It is therefore extremely important to insist that launchers have engines capable of reigniting, allowing them to guide the re-entry away from populated areas, usually in a remote area of the ocean.

As already mentioned, performing a controlled re-entry also requires the presence of additional fuel on board, on top to that needed to launch the payload. Some launch service providers, who use modern rockets with reignitable engines, exhaust the onboard fuel to push the payload as high as possible, thus saving time and fuel for customers, as otherwise the payload would have to use its own thrusters to raise the orbit.

Deorbiting the launcher in a controlled manner involves costs that not everyone is willing to bear, also because the risks of accidents are usually evaluated on a launch by launch basis, thus keeping risks at level low enough to make it easier for governments to justify uncontrolled re-entries. However, with the progressive increase in the number of launches carried out each year, it is necessary to also consider cumulative risks. The challenge, in an increasingly diversified and competitive space launch market, is not only to raise safety standards, but also to ensure that everyone is subject to them, so as not to penalize anyone.

In this landscape, characterize by a lack of international safety regulations, it is pleasing to note that the new European launcher Ariane 6 has the potential to be more ecological because the engine of the second stage, called Vinci, has been designed to be ignited and turned off multiple times. This will allow the stage to deposit satellites at different heights and then "deorbit" and re-enter the Earth's atmosphere, after completing its tasks, rather than contributing to the growing problem of space debris. Unfortunately, during the first launch of Ariane 6, which took place on July 9, 2024, the Vinci engine shut off during the second ignition and refused to reignite, turning the second stage into a space wreck that will fall uncontrollably. A real shame but we hope that the problem will be solved and the next launch, scheduled for the first quarter of 2025, can perform the controlled re-entry.

To get an idea of the number of re-entries (controlled and uncontrolled) that occur above our heads, I recommend a visit to the site https://aerospace.org/reentries?page=1.

Also in the re-entering objects, we see the growing contribution of Starlink satellites. While, in the last week of September 2024 (from Sept 23 to 29), 12 objects re-entered, 8 of which were Starlink satellites, the remaining 4 being 2 satellites, 1 rocket body and the "trunk" of the Polaris Dawn capsule, in the first week of February 2025 (from Feb. 1 to Feb. 7) of the 36 objects which re-entered all but 3 were Starlink satellites.

What happens to all this material? We know that upon re-entry, satellites are destroyed by friction, but it would be wrong to think that they disappear without leaving a trace. The overheated material turns into toxic gas and dust which will remain suspended in our atmosphere, perhaps also scattering solar radiation and contributing to climate change.

Using NASA's WB-57 jet, a special plane capable of flying in the stratosphere at 19 km altitude (roughly twice the altitude of commercial aircraft) that took off from FairBanks in March 2023, samples were collected over Alaska and much of the United States at an altitude where the only natural source of metals is from meteorites that vaporize due to friction with the atmosphere. This is about 40,000 tons of celestial material per year, equivalent to about 100 tons per day, mostly dust and fragments of carbonaceous rocks. Metallic meteorites are a small minority in space and consist almost entirely of iron and nickel. There are significant differences between the vaporization of meteorites and that of satellites. While most of the meteoric mass (originating from a large number of submillimeter objects) is deposited at altitudes between 75 and 110 km, spacecraft, which are larger and move slower than meteorites, vaporize between 40 and 70 km over a region about 300 km long. On a yearly basis, we can count several hundred large re-entry events of satellites and upper stages, and we know that every one deposits up to several tons of mass compared to the micrograms of individual meteorite.

Atmospheric circulation drags the spacecraft's generated particles, as well as those of meteoric origin, to lower altitudes in the stratosphere where mixing and coagulation in the aerosol particles occur.

The analysis of the samples revealed that the stratosphere is already scattered with metals from the re-entry of spacecraft. While almost all stratospheric particles also contain meteoric metals, aluminum, silver, niobium and hafnium, for example, are not found in meteorites but are used in spacecraft. Niobium and hafnium are markers of the re-entry of some rocket nozzles. It is hypothesized that silver may be a marker of electronics that is relatively more common in satellites compared to the stages of exhausted rockets. The aluminum from the spacecraft is generally found in the same particles as niobium and hafnium, specific to rocket engines.

It is noteworthy that the products of the re-entry of vaporized spacecraft above 50 km altitude can be measured with such sensitivity in aerosol particles at less than 19 km altitude. Hafnium is not only detectable but also quantifiable compared to niobium. Yet hafnium is an element used in only one component (<1% of the mass) of some types of launchers.

However, these metals, along with aluminum, were included in about 10% of the most common particles in the stratosphere.

Of course, this percentage can only increase with the increase in the number of launches and re-entries of satellites.

What could be the effect of all these metallic dusts? Sierra Solter, a plasma physicist, argues that all these conductive materials, accumulating in the ionosphere and in the magnetosphere, could act as a magnetic shield, potentially weakening or deflecting the Earth's magnetic field. While her article has aroused much interest and as many criticisms, we are all aware that life on earth is possible precisely thanks to the magnetic field that protects us from the harshness of the Sun.

It is too early to say how these metals will affect the stratosphere and the ionosphere, but having data is the first step to assess the environmental implications and the need for regulations to reduce the impact of the space industry. If the data show that a safety threshold is being exceeded, it might be advisable to limit the number of launches or to ask industries to use new materials.

In fact, the materials do not necessarily have to be "new". Inspired by ancient traditions, in Japan there are plans of replacing some metal parts with wood. Researchers at the University of Kyoto have conducted tests with various types of wood to determine how well they could withstand the rigors of launching into space and of long flights in orbit around the Earth. After laboratory tests in which the wood samples did not undergo measurable changes in mass nor showed signs of decomposition or damage, the samples were sent to the ISS, where they were subjected to exposure tests for almost a year before being returned to Earth. Even in this case, no obvious signs of damage were found, a phenomenon that can be attributed to the fact that in space there is no oxygen that could burn the wood, and no living creature that could cause it to rot. The best wood that passed all tests, resisting cracking, was magnolia, which was used for the LignoSat probe, built by researchers at the University of Kyoto and by Sumitomo Forestry company. The satellite, which is a cubesat equipped with a series of instruments that will measure the vehicle's performance in orbit, reached the ISS in November 2024 and in December was deployed into orbit from the Japanese Kibo module.

After being operational for 6 months, LignoSat will re-enter the atmosphere and test the idea of using biodegradable materials like wood, to see if they can represent a more environmentally friendly alternative to the metals with which all satellites are currently built.

Lunar Ecology in the Era of Commercial Exploitation

After being visited by dozens of Soviet and American missions between 1959 and 1976, the Moon went out of fashion. The Apollo missions had sanctioned the American victory in the space race and, once the opponent was beaten, NASA's attention (and that of US Congress, which saw no reason to fund costly human expeditions) turned to other objectives with sporadic scientific missions to continue studying the Moon from orbit. Not that there were no supporters, even authoritative ones, for the return to the Moon, but politics did not show sensitivity to the cause. The Apollo program taught us that great enterprises require large funding over a time frame much longer than the duration of a presidency, so it requires solid political will. Instead, NASA and Congress have oscillated according to the preferences of the succeeding Presidents. George W. Bush began talking about returning to the Moon in 2004 without however providing sufficient financial coverage. In 2008 the President changed and Barak Obama proved more sensitive to supporters of Mars and asteroid exploration. After all, the United States had already been to the Moon. Obama left his mark on the history of US space thanks to the idea of

P. Caraveo, *Space Ecology*, https://doi.org/10.1007/978-3-031-78344-9_5

revolutionizing the way of sending humans and materials into orbit by opening the business to private companies, no longer executors of NASA projects but sponsored by the Agency to develop their own technological solutions. It is thanks to this new political vision that SpaceX has become such an important player in the American and global space arena. Trump's inauguration in 2016 changed the wind and NASA was ordered to prepare plans for the return to the Moon with some haste, given that, thinking about re-election (which did not happen) the President asked to bring American astronauts back to the Moon by 2024. Thus, NASA began the program called ARTEMIS, a doubly apt name since it is about the twin sister of Apollo, goddess of the Moon, who baptizes the program that will bring the first woman (and the first non-white man) to our satellite. Trump's lunar choice, later confirmed by Biden, was probably inspired by the very clear Chinese ambitions that, in the new millennium, have become important and very successful players on the lunar chessboard. However, while the Artemis program is progressing, the political wind may have changed yet again. During his Inaugural Address, in Jan. 2025, President Trump promised to launch American astronauts to Mars.

The Lunar Renaissance and Its Diplomatic Consequences

In fact, the revival of interest in our satellite has occurred also thanks to the expansion of the audience of space actors. Alongside the ever-increasing number of space agencies wanting to reach the Moon, with missions in orbit and on the ground, a whole new world of private entrepreneurship has opened up, which, with the pursuit of space profit, could give new impetus to the return of human explorers to the Moon.

Starting with the activities of the space agencies, in addition to NASA (which has started to collaborate with private industries), the European Space Agency (ESA), the Japanese one (JAXA), the Indian one (ISRO), the Russian one (Roscosmos) and the Chinese one (CSA) have entered the scene. The three Asian actors have a growing lunar program: all start with missions in orbit, to master the technique of insertion into lunar orbit, with the aim of subsequently moving on to missions that foresee landing on the moon. This is a maneuver, successfully executed in the '60s and '70s, which the public tends to give for granted, but which is anything but trivial and can reserve unpleasant surprises.

If we look at the history of the race to the Moon, in the '60s, when the United States and the Soviet Union were challenging each other in the space adventure, we realize that every success was preceded by several failures. The soft landing, achieved in 1966 first by the Soviets and then by the Americans, came after at least a dozen failures that had equally befallen the two superpowers. Anyone who has doubts should reflect on the fact that the grand total of lunar probes, from the first attempts in the late '50s to Apollo 11, was 72 missions, only 25 of which were successful. As easily understandable, the failures were concentrated in the initial phases of the Soviet and American programs, when it was necessary to learn how to perform all the maneuvers, but one should never take anything for granted: having successfully completed a difficult maneuver is not an insurance for the future.

The first space power to land on the moon in the new millennium was China, under the auspices of Chang'e, the moon goddess of Chinese mythology. After the preparatory missions Chang'e 1 and 2, which orbited around the Moon in 2007 and 2010, the Chang'e 3 probe landed on the lunar surface in 2013, 37 years after the last Soviet mission Luna 24 (launched on August 9, 1976 and landed in the Sea of Crises where it collected, and brought back to Earth, 170 g of material) and 41 years after Apollo 17.

The success of Chang'e 3 gave China an important space primacy. They managed to successfully complete the landing maneuver at the first attempt, avoiding the failures that half a century earlier had hit the Soviet Union and the United States. The success of Chang'e 3's landing, which carried the Yutu (jade rabbit) rover for the exploration of the surrounding terrain, represented the first step towards a more difficult goal no one had ever attempted: landing on the far side of our satellite. This was the task of Chang'e 4, which landed in January 2019 in the Von Karman crater. Since the probe and its Yutu 2 rover could not be seen from Earth, the Chinese space agency, in May 2018, had launched beyond the Moon the Queqiao satellite which had the task of acting as a radio bridge. Chang'e 4 was the first example of a mini lunar greenhouse where cotton seeds sprouted. Unfortunately, the cultivation had a short life because the frost of the lunar night froze the seedlings, but the experiment was a milestone.

In the course of 2019, the same luck did not favour a private Israeli company nor the Indian space agency, whose probes failed in the final stages of the landing maneuver. On April 11, 2019, Beresheet (genesis, in Hebrew), the first private Israeli mission that had no commercial purpose, but rather aspired to be a cultural mission, crashed. Funded by Morris Kahn, an Israeli billionaire, it wanted to bring to the Moon a time capsule with 30 million pages of data containing the entire Wikipedia in English, the Torah, children's drawings, testimonies from Holocaust survivors, a copy of the declaration of independence, the flag and anthem of the state of Israel. Then at the last moment, tiny tardigrades were also added (the most resilient living beings known) encapsulated inside layers of resin, a decision heavily criticized because it goes against the rules of planetary defense, as we will see later. It should be noted that the private Israeli mission inaugurated a new "rapid" approach to moon landing since it did not follow the progression of having an instrument in orbit first and only later a moon landing maneuver. Rather, they wanted to skip the first step to move directly to the second. A strategy dictated by the need to maximize the economic return, or, in the case of Beresheet, of image, for the private company and for their financiers. A risky strategy that has not yielded the hoped-for results even though the failure of Beresheet's "rapid" approach was followed by that of the more traditional path followed by the Indian Space Agency.

In fact, on September 6 of the same year, the lander Vikran of the Indian mission Chandrayaan-2 crashed, forcing the prime minister Narendra Modi, who was present in the control room during the moon landing, to confort the mission leaders who were crying in despair. "We got very close but we will need to go further in our next attempts".

Certainly, the Indian and Israeli missions were carried out at low-cost, with an investment of 150 million dollars for Chandrayaan -2 and about 100 for Beresheet, significantly lower than that expected for comparable missions carried out by NASA or ESA.

On the other hand, the Chinese program continued to enjoy good luck and, in December 2020, successfully completed Chang'e 5, the first automatic sample collection mission. After a brief period of stasis, in 2023, the lunar traffic began to grow again. The private Japanese mission Hakuto-R, which wanted to deposit two small commercial payloads on the Moon was the first. On April 25, 2023, it seemed that the moon landing maneuver was proceeding smoothly but then, when the probe was very close to the lunar surface, contact was lost. The Japanese control center stated that it was unable to confirm the success of the mission that was not supposed to have made a soft landing.

Perhaps the engines, which are absolutely necessary because the Moon does not have an atmosphere and parachutes cannot be used, did not work, or they turned off too early due to lack of fuel or a malfunction of the onboard software that misinterprets the data it receives and, believing it has reached its destination, orders the engine shutdown.

As it was the case for Beresheet, also Hakuto-R, after being launched with a Falcon 9, had successfully completed all the required maneuvers, entering lunar orbit, but failed the final test.

Both teams knew they wanted to do something very ambitious. If they had succeeded, they would have been the first private company to land on the moon, and they would have done it during their first mission, improving, if possible, the success of China which had indeed landed on the moon on the first attempt, but after two missions in orbit.

August 2023 gave us new lunar emotions. Determined to seek revenge, on July 14, 2023, the Indian Space Agency (ISRO) launched a revised and corrected replica of its previous mission calling it Chandrayaan 3. Also in this case, the probe had to follow a spiral orbit designed to minimize fuel consumption while lengthening the transit time to over a month, with a moon landing scheduled for August 23.

Just as Chandrayaan 3 was describing its increasingly larger elliptical orbits, the Russian space agency ROSCOSMOS announced that the Luna 25 probe was ready for a launch scheduled for August 11. The date was not chosen at random, as it fell less than a day before the

47th anniversary of Luna 24, the last successful mission of the Soviet lunar program. It was natural to name the new mission Luna 25, to underline the continuity with a tradition of automatic missions that, despite struggling to compete with the visibility of American astronauts, opened important chapters in the exploration of the lunar surface. In the case of Luna 25, the direct orbit would have been followed and, although the Russian probe had left after the Indian one, the moon landing was expected a couple of days in advance.

Unfortunately, the Russian lunar program must have rusted a bit and the landing maneuver was not executed correctly. Luna 25 crashed, creating a new small crater on the surface of the Moon. So what was supposed to be a triumphant occasion for Russia on the international scene became an embarrassing fiasco.

The desire to arrive first perhaps did not help, but the Russian failure certainly raised the level of attention on Chadrayaan-3.

The trepidation in India was palpable, since, in 2019, everything had gone well until the fatal malfunction of the landing software. The fascinating Moon does not forgive even the smallest mistake. Turn off the engines too early, as happened to the Japanese lander a few months before, to the Israeli one and to Chandrayaan 2 in 2019, and the probe touches the ground at too high a speed and is destroyed. On the other hand, keep the engines on too long during the preliminary maneuvers for the modification of the orbit in preparation for the descent, as happened to Luna 25, and the probe is put on the wrong trajectory that ends with a disastrous impact. It seems that the Moon wants to remind us that nothing is trivial in a space mission: having managed to land on the moon in the past is not enough, what counts is what happens in the last crucial minutes.

On August 23, all of India participated in the moon landing of the Chandrayaan 3 mission and national pride soared, indeed on the Moon because India became the fourth space power to perform a Moon landing.

For India, it was the second attempt and perseverance paid off: the moon landing, broadcast live from the control room, was perfect. Prime Minister Modi, connected from South Africa where he was attending the BRICS congress, expressed all his satisfaction and said that it was not just an Indian success but of all humanity. The fact that the moon landing was scheduled just in conjunction with an important international event was certainly not a coincidence. It is a careful planned choice that makes clear the political and strategic value that the Indian government attributes to space missions.

Once landed, the Vikram lander, which is the size of an SUV, released the small 26 kg Pragyan rover which explored the surroundings for 2 weeks, during the illuminated period of the lunar day. Then the mission was hibernated in the hope that it could survive the freezing lunar night. Unfortunately, at the return of the Sun, which should have charged the batteries, there was no response to the messages from Earth. In fact, it was expected that the mission would be operational only for one lunar day, but many hoped it could continue.

They are not exactly in the lunar south pole, but, compared to all the other missions that have touched the surface of the Moon, their landing site, at 70° latitude south, is definitely the closest to the south pole. They are certainly the first to approach the lunar eldorado where the presence of ice opens interesting scenarios for future human colonies and for space business. This also goes into the account of Indian space pride.

Then, on September 6, the Japanese space agency launched the SLIM mission (for Smart Lander for Investigating the Moon), which has followed the low fuel consumption trajectory to arrive in lunar orbit in a few months, in view of a moon landing at the end of January 2024.

But lunar traffic is also increasing thanks to the NASA Commercial Lunar Payload Services (CLPS), created with the aim of making trips to the Moon more efficient by entrusting private companies with the task of providing transport services to send equipment, landers and rovers. With the CLPS program, NASA wants to replicate on the Moon the commercial policy it has followed for Earth orbits where it has entrusted the launch of satellites, but also all the traffic of goods and people to the International Space Station, to private companies that have been funded to develop launchers and capsules and now have contracts to provide launch services to the Agency.

2024 lunar exploration started with the launch of the Astrobotic's Peregrine mission, which aimed to be the first American commercial mission to land on the moon with its NOVA-C lander. Unfortunately, also Peregrine was not blessed by luck: due to a fuel leak, it could not even attempt the insertion into lunar orbit and sadly returned to destroy itself in the Earth's atmosphere. However, looking at the mission's budget, one begins to understand the new market of lunar logistics, where transport plays the lion's share. In fact, in the case of Peregrine, NASA had funded five instruments for an amount of 9 million dollars while transport cost 108 million. Even if these may seem like significant figures, we are talking about low-cost lunar missions and it was understood that some could go wrong.

To lift the spirits of lunar exploration enthusiasts, the SLIM mission of the Japanese space agency, on January 20th, made a soft landing, making Japan the fifth nation to have succeeded in the maneuver. The SLIM mission teaches that it is still possible to achieve some "first times" on the Moon since it can claim at least two; on one hand it can boast a very precise landing having landed just 50 m from the expected point, on the other it is the first probe to have gently touched the lunar surface upside down, as shown by the photo taken by one of the two small rovers that detached before the probe touched the ground (Fig. 1).

Fig. 1 SLIM at the top the graphic representation of the satellite in the correct position for landing, at the bottom the photo of the upside-down landing. (Credit JAXA)

The unusual landing prevented the solar panels from being at the right angle to receive the sun's rays and the probe, without energy, was put into hibernation. It was hoped that the path of the Sun during the lunar day (which lasts 2 weeks) would have come to illuminate the panels, allowing the probe to be restarted, which regularly happened on January 28, 2024. The dead and resurrected probe is another lunar first since, in general, turned off probes do not survive the cold of the lunar night. For SLIM it was different also because it was in broad daylight in less extreme temperature conditions. However, the probe astonished its builders since it survived its first lunar night, contrary to all predictions. When the Sun returned to provide energy, the probe called home, surprising the control center that was about to send the staff home since everyone thought the mission was over. Indeed, SLIM survived few more lunar nights until the mission was terminated at the end of August. It remains true that, while the precision landing was one of the key points of the mission that had to land less than 100 m from the point pre-established, the other two first times happened by chance.

Then it was the turn of a new mission from the CLPS program with the challenging name of Odysseus, the first private American mission to land near the south pole on February 23, 2024. In fact, not everything went as planned and Odysseus, after touching the lunar surface, settled on its side. The unusual horizontal position, with the antennas facing the ground, caused serious difficulties in sending data to Earth. Despite being able to operate only for a week, Odysseus represented a milestone in the history of space exploration: being the first "private" mission to land on the Moon in the strategic region of the South Pole gave an enviable primacy to the company Intuitive Machines.

That's what NASA wanted since, as the main customer, it paid 118 million dollars for the transport of 6 instruments worth 11 million. Odysseus also carried instruments provided by other customers, including a compendium of 125 mini sculptures of the Moon created by the sculptor Jeff Koons, who pointed out that this is the first "authorized" work of art to land on the Moon (Fig. 2).

Due to the skewed landing, which the sculptures' box directly in contact with the lunar surface, probably embedded into the lunar regolith, not all the instruments worked as expected, but the mission was archived as a success. After all, Homer would have been happy to know that near the lunar south pole there is a lander called Odysseus that brought the United States back to the Moon after 52 years from the end of the Apollo program.

Three lunar missions in the first 2 months of 2024 testify the renewed interest in exploring our satellite that

Fig. 2 Moon sculptures attached to the Odysseus lander. (Credit https://jeffkoonsmoonphase.com)

continues with Chang'e 6, a Chinese mission, launched on May 3, 2024 with the goal of landing on the far side of the Moon to collect samples to bring back to Earth.

The mission entered lunar orbit on May 8 and performed orbit corrections while waiting for optimal lighting conditions at the chosen landing site. This is the South Pole-Aitken basin, one of the largest in the solar system with its 2000 km in diameter and 8 km in depth. It is a place geologically very interesting because it is expected that the impact responsible for the creation of the basin dug material from the lunar mantle.

The Chang'e-6 lander, equipped with a camera, a spectrometer and a radar to study the surrounding environment and choose the point where to collect a sample, landed gently on June 1 and activated its mechanical arm to collect both soil and subsurface samples up to 2.5 m deep with a drill. Chang'e 6 also brought 4 international instruments to the Moon: a French Radon detector, an Italian laser retroreflector, a Swedish instrument to measure negative ions, and a Pakistani cubesat (Fig. 3).

Fig. 3 The Chang'e 6 lander on the Moon photographed by the rover. (Credit CNSA)

The samples collected over 2 days of activity were inserted into the ascent vehicle that lifted off from the Moon on June 3, delivering the material to the orbiter where it was transferred to the return module for its journey back to Earth.

Since it is impossible to establish direct communications with the hidden side of the Moon, the Chinese space agency has launched a satellite to act as a repeater and connect the lander with the control center. Following the procedure used in 2018, when the Queqiao satellite was launched into lunar orbit to transmit information from Chang'e-4 to Earth, Queqiao-2 was launched in March. The pair was used in tandem to stay in contact with Chang'e-6 during the sample collection.

On June 25, the return module landed in Inner Mongolia bringing back to Earth, for the first time in history, samples from the far side of the Moon that Earth never sees.

More ambitious U.S. commercial missions were planned during 2024. At the end of the year, NASA's VIPER rover was supposed to rest in the lunar South Pole region, where it was to search for water ice. In mid-July, however, NASA stunned everyone by announcing the cancellation of the VIPER probe. It is hard to understand why the agency that has already invested $450 million in the VIPER rover, whose construction is completed, and has a $323 million contract with Astrobotic to transport it to its destination made such a drastic decision. What's more, since the contract with Astrobotic cannot be canceled, there is a risk of having a model with the same mass as the rover launched unless someone has an instrument, of the right size and weight, ready to launch. Inded, Astrobotic received 60 inquiries and, at the beginning of February 2025, announced the selection of FLIP (short for FLEX Lunar Innovation Platform) a 1000 pound rover by Venturi Astrolab Inc.

FLIP is roughly the same size of VIPER and plans to test technologies to minimize problems caused by particles of lunar dust. NASA said that VIPER's cancellation does not change its interest in precisely locating and estimating the amount of ice accumulated in the ever-shadowed South Pole craters, a key step the begin building a permanent lunar base. On January,15 Falcon9 launched Blue Ghost, a robotic lander that Firefly has developed to take scientific instruments to the Moon. Blue Ghost carries 10 NASA payloads, including an X-ray camera, a drill to measure the flow of heat from the Moon interior toward the surface and an electrodynamic dust shield to clean off glass and radiation surface. Through this and other missions in the commercial program, NASA wants to continue to pave the way for the return of human beings to the Moon. In mid 2027 the Artemis III mission will bring the first woman and the first non-white man to walk on the Moon. For permanent settlements we will have to wait a little longer, but their location will be decided on the basis of the golden rule of the real estate market: what counts is mainly the position. It is no coincidence that next year also the Chinese mission Chang'e 7 will be directed to an adjacent area, always in the South Pole of the Moon, because everyone wants to be close to the much coveted ice deposits and in areas that enjoy almost constant illumination to allow continuous use of solar panels to provide energy to the bases.

The supply of energy, in fact, is a delicate problem for future lunar colonies which, unless they are in strategic positions in polar areas perpetually illuminated by the Sun, must learn to live with the scorching lunar day lasting 2 weeks, followed by 2 weeks of freezing night. Storing energy to survive 2 weeks of intense cold is not trivial also because batteries are heavy and you can't certainly imagine transporting them in quantity sufficient for a human settlement.

The idea is rather to exploit the hydrogen contained in the lunar ice. We need to invent solar-powered robots capable of collecting it and then electrolytically decompose it into hydrogen and oxygen, two gases that can have multiple uses. While oxygen could be used by the settlers (who need to breathe), hydrogen could provide energy during the freezing lunar night thanks to the use of fuel cells.

In parallel, there are those who think of turning the Moon into a "gas" station for interplanetary probes that could leave Earth with a half-empty tank (thus lighter), and then refuel with hydrogen and oxygen on the Moon. Sure, lunar fuel would be more expensive than terrestrial one, but the savings on launch costs would balance the accounts. Water is the gasoline of space.

For this reason, the plans for lunar settlements are inextricably linked to the exploitation of ice which, however, is a resource available only in the perpetually shadowed polar craters.

The situation is not simple because everyone wants to settle in the same area that must be occupied as soon as possible to prevent others from settling there. Since, according to the treaty on the peaceful use of space signed in 1967, no state can claim ownership of celestial bodies, one should ensure the right of use. In fact, until the area is occupied, it will not be possible for others to install there. For this reason, it is necessary to act in advance to avoid conflicts or, even worse, to prevent the moon from becoming a cosmic Wild West.

As described in the Appendix, there is a treaty on the Moon that would provide a regulatory framework for the new exploration, which, however, is of no use because it has not been signed by the main space powers.

The matter was taken up again in the ARTEMIS agreements (summarized in the Appendix) signed by over 50 nations. However, we are not talking about an international treaty: the agreements are only valid for the States that have signed them. In the new race to the Moon, on one hand we find NASA and all the nations that have signed the Artemis agreements, on the other Russia and China along with the nations that collaborate with them.

And everyone wants to occupy the best place in the South Pole lunar area where the presence of ice is both a scientific and commercial magnet. In fact, scientists believe that the lunar ice, certainly very ancient, has extraordinary potential to help us understand parts still unclear of Moon's history (and by extension that of the Earth). On the other hand, there are private investors who, strengthened by the U.S. Commercial Space Launch Competitiveness Act, crafted by the Obama administration in 2015 precisely to encourage private investments, feel authorized to proceed in the exploitation of lunar resources. As explained in the Appendix, they will not be able to declare themselves owners of the area of their interest but they can occupy and commercially exploit it. Will they respect the fragile lunar ecology? How will they react if other investors or other nations want to exploit the same deposits?

Space diplomacy is more necessary than ever.

Ecology of Lunar Settlements

The settlements will have to be built and managed with long-term sustainability strategies that imply respect for the delicate ecosystem of the Moon, use of local material and meticulous recycling. In fact, the word sustainability is the mantra of the new space exploration. A mantra that must be applied both to settlements and ground activities, as well as to everything that orbits around the Moon or that wanders in the Earth-Moon system risking to impact on the lunar surface.

Let's start considering the impact of any type of activity on the lunar environment.

The experience of the Apollo program teaches us how difficult it is to work with the constant presence of regolith. The "Moon dust" is gray, fine, abrasive and electrically charged, made up of silicates with an average size of 70 microns. These grains were created over billions of years, by the impacts of meteorites and asteroids that have reduced much of the lunar soil to dust. The absence of an atmosphere also means that wind and water erosion (common on Earth) is absent. Finally, the constant exposure to the solar wind has left the lunar regolith electrostatically charged, which means it adheres to everything. We saw this on the suits of the astronauts who came out of the lunar modules immaculate only to appear gray upon return. And the dust did not just settle on the suits, once the regolith made its entrance into the modules it stuck to everything and became a health hazard, causing eye irritation and breathing difficulties. The problem will become increasingly important when humans and robots will perform longer and more complex activities on the lunar surface. In a base that needs to be supplied with material and personnel there will be the problem of take-offs and landings of vehicles for the transport of things and people that will raise clouds of regolith that will float for a long time before settling again. The Apollo landing sites took years to return to the original situation. But today we are talking about sites that will continue to be inhabited after the release of the regolith clouds which, in addition to being dangerous for the spacecraft, that might not have a good view of the landing site, can damage the nearby equipment, disturb sensors, clog mechanical gears and degrade optical surfaces or solar panels. These are not hypotheses, in November 1969 the exhaust jets of the engine of the lunar module (LM) of Apollo 12 reached the Surveyor 3 spacecraft, located 160 m away. We know this because astronauts Pete Conrad and Alan Bean in November 1969 inspected this unmanned vehicle that had been sent in 1967 to explore the region of Mare Cognitum with the task to characterize the lunar soil in view of the crewed missions.

The Apollo 15 mission that landed in the Hadley-Appennine region in 1971 provides another example. During the descent of the LM, the astronauts David R. Scott and James B. Irwin were unable to see the planned landing site because the exhaust gases had raised a thick cloud of regolith. This forced the crew to choose a new, nearby landing site.

Two historical events to demonstrate how the regolith kicked-up by gas jets can become a danger, especially when other spacecraft and structures are nearby.

While it is true that the regolith did not compromise the execution of the Apollo missions that landed in the equatorial or mid- latitude regions, future Artemis (and CLPS) missions will take place at the lunar South Pole, where the soil is presumed to be significantly more porous. Moreover, to meet the logistics of a lunar settlement, a large number of landings and takeoffs are expected each of which will have payloads much larger than those of Apollo, requiring a greater thrust to slow the descent. Consequently, the exhaust gases could generate craters by spraying the regolith in all directions, creating potential hazardous conditions for the Artemis base camp, composed of fixed living modules, a habitable mobility platform, a lunar terrain vehicle (LTV). Therefore, to enable a "sustainable program of exploration and lunar development" it is necessary to protect humans, structures and vehicles from the risks of regolith that, traveling with the astronauts, could also affect the operation of the Lunar Gateway.

Unfortunately, experimental investigations on lunar regolith are extremely difficult because lunar conditions are very different from those on Earth. The lower gravity (about 16.5% of Earth's), the vacuum and extreme temperature variations are not reproducible in the laboratory and therefore simulations must be used.

Since fences around the landing zone are not capable of satisfactorily confining the regolith, different approaches such as landing platforms built by the probes themselves, spraying molten aluminum, or treating the regolith with microwaves to transform it into ceramics are being considered.

Regolith is not just a nuisance, it is the only material readily available on the Moon and, once compacted using laser, it could be used to build solid screens to protect astronauts from cosmic radiation, but also from all the material that naturally rains from the sky. Without a magnetic field and without an atmosphere, the physical protection of astronauts is of vital importance.

While the Apollo project has taught us how long it takes for lunar dust, stirred up by astronaut activities and from probe landings and takeoffs, to settle, to assess the possible environmental consequences of major soil disturbance we have examples of both planned impacts and cosmic surprises.

What Happens to Lunar Probes? Ecology of Lunar Orbits

In the census of terrestrial artifacts resting on the Moon (https://en.wikipedia.org/wiki/List_of_artificial_objects_on_the_Moon) we find the remains of Soviet, American and Chinese missions that have landed, the two Soviet Lunakods and the Chinese rovers, 3 electric cars that transported the astronauts of the Apollo 15, 16, 17 missions, the instrumentation deposited by the Apollo missions along with the legs of the lunar modules (which served as launch pads).

We must then add the remnants of the impacts of the first lunar missions (which were only meant to hit the Moon) and those that marked the inglorious end of failed landing maneuvers (in 2019 for the Israeli mission Beresheet and the Indian Chandrayaan 2 and in 2023 the failed Japanese and Russian missions, along with the intact lander of Chandrayaan 3 and what has landed and will land from 2024 onwards (so far SLIM, Odysseus and Chang'e 6). But the list of missions that have reached the lunar surface represents only a fraction of the 200 tons of junk we have deposited on the Moon.

To understand the terms of the problem, we must remember that all probes that enter lunar orbit, or that land on the Moon, do not arrive "alone". Rather they are accompanied by a part of the rocket that, after launching them, provided them with the additional thrust necessary to reach escape velocity from Earth gravity to begin the Moon journey. Even if the last stage of the rocket detaches, it continues by inertia to move at the same speed and with the same trajectory as the probe that arrives at its destination with at least one accompanying piece that then continues to orbit around the Moon. In the case of landing missions, it is possible that a part of the probe remains in lunar orbit, while the rest performs the maneuver to gently land on the surface. The Apollo missions, for example, in addition to the third stage of the Saturn 5 that accompanied them on the journey, left the lunar module in orbit, which, once the astronauts were back to the command module, had completed its task. The effect of the impact of the lunar modules (and sometimes the remnants of the carrier rockets) was measured by the seismographs that the astronauts had left on the surface, to study the internal structure of the Moon using the propagation of seismic waves.

The category of impacts also includes those planned at the end of a mission in order to use a piece of space junk to produce a cloud of debris to be analyzed from Earth, to investigate for example the presence of ice. This is what was done on October 9, 2009 at the end of the LCROSS (for Lunar Crater Observation and Sensing Satellite) mission which crashed into the Cabeus crater near the lunar south pole. In fact, the last stage of the Centaur launcher arrived first, and its impact was seen by LCROSS before it was its turn to crash. The multiple impact was recorded by the Lunar Reconnaissance Orbiter (LRO) and by ground observers.

After this long preamble, it will be clear that what remains of a lunar mission will end up producing a new crater on the surface of our satellite.

However, no one had considered the possibility that a piece of interplanetary garbage could hit the far side of the Moon, as happened on March 4, 2022. The culprit was clearly the left-over of a mission flown in past years. When Bill Gray realized that there was something describing a trajectory destined to hit the Moon, the question "from which mission does it come from?" arose spontaneously.

Asking to whom a piece of space debris belongs is a non-trivial question that goes beyond curiosity. Considering that, the 1972 Convention on Liability (see Appendix) states that the nation that carried out the launch is responsible for the damages caused by any component of a space mission, it is important to identify the owner of the debris.

At first, he thought it was the last stage of the Falcon9, which, in 2015, had taken the NASA DSCOVR mission to L1 (which is located 1.5 million km from Earth, about 4 times the distance of the Moon). However, a more accurate examination of the trajectory, and an exchange with experts from JPL, showed that it could not belong to that mission. By process of elimination, he then thought that it was a piece of the Long March 3C launcher from the Chang'e 5 T1 technology mission, which China had launched in 2014. It was part of the preparation for the Chang'e 5 mission, which, in December 2020, collected and returned lunar soil samples to Earth. The mission in question had flown by the Moon and then returned home because it had to test the technique of re-entry into the atmosphere.

Although there is nothing and no one on the far side of the Moon who can claim to have suffered damage, to avoid misunderstandings, the Chinese space agency denied that it was something of their property, saying that the last stage of that launcher had been destroyed re-entering the atmosphere. However, the Space Force's 18th Space Control Squadron (18SPCS) which monitors the plethora of space debris orbiting Earth questioned this claim. According to their data, the launcher of the Chang'e 5 T1 mission, which was on its way back to Earth, did not re-enter the atmosphere, so it must have spent the last 8 years traveling between Earth and the Moon, following complicated trajectories due to the continuous interaction with the gravity of the two celestial bodies, until it entered a collision course. The official statement is as follows:

While U.S. Space Command can confirm the CHANG'E 5-T1 rocket body never de-orbited, we cannot confirm the country of origin of the rocket body that may impact the moon.

If the impact is due to part of the Chang'e 5 T1 mission launcher (as stated on the Wikipedia page that lists the artifacts on the Moon https://en.wikipedia.org/wiki/List_of_artificial_objects_on_the_Moon), it is worth noting that the small 4M mission for Manfred Memorial Moon Mission managed by LuxSpace, a Luxembourg company connected to the German OHB System, was traveling along with the launcher debris. The 4M mission wanted to honor Manfred Fuchs, the founder of OHB, who died in 2014. After the lunar flyby on October 28, 2014, which made LuxSpace the first private company to perform one, 4M continued to transmit for about a month, while traveling with the rest of the launcher. Thanks to the 4M mission, Luxembourg could be the eighth nation to leave its mark on the Moon after USSR, USA, ESA, Japan, India, China, and Israel.

However, the mysterious impactor has sounded an alarm bell reminding us that no one is cataloging the many objects (some say even 200) that orbit around the Moon. A problem that perhaps should be considered, given the renewed interest in our satellite that will see no less than 50 missions in the next 5 years https://en.wikipedia.org/wiki/List_of_missions_to_the_Moon. No one wants to run the risk of receiving an unplanned visit, perhaps on a lunar base that hosts human crews. Let's not forget that the Moon cannot count on the help of the atmosphere which, on Earth, welcomes debris in a fiery way, destroying them before they cause damage.

The cosmic vacuum lets everything pass and the crash is inevitable. In addition to the dangers for ongoing activities, we must also avoid that the debris goes to ruin the iconic places of the Apollo mission landings, which are considered akin to archaeological sites and, as such, should be protected.

Interestingly, even though everyone agrees with the idea, it is not easy to ask that the sites be recognized as UNESCO World Heritage Sites because no institution can submit the application. In fact, since the treaty for the peaceful use of Space prohibits any nation from declaring its sovereignty over a celestial body or part of it, no state can claim ownership of the places for which protection would be sought. Sure, NASA owns the instrumentation and everything that remained on the surface of the Moon, but that is not American territory. Since the application for a site to be considered by UNESCO as heritage of Humanity must come from the government of the state where the site is located, we are in a legal dead end. Since the Moon is not owned by anyone, no one can ask for a part of it to become a heritage of Humanity.

We certainly do not want these historical sites to be defaced, neither by unplanned impacts nor by commercial missions hunting for souvenirs. The photos from the LRO mission show us that the Moon is very good at preserving historical memory. Unpleasant interferences must be avoided.

Let's Look at Mars

The summary infographic of the exploration of the bodies of the solar system, although not very up-to-date because it was made in 2012, makes it clear that Mars is the most visited planet after the Moon (Fig. 1).

The color code (which distinguishes successful missions from those failed) also immediately makes us appreciate how the different celestial bodies have received attention (thus have been visited) by different space powers. The Moon was a challenge between the USSR and the USA until China entered the field in the new millennium. Instead, Mars has been a very American affair and is poised to become ever more so judging from the statements from President Trump.

The rovers that have explored and are exploring Mars are American, and only recently, China has successfully performed the maneuver of landing, a maneuver that has unfortunately not yet been successful for the Europeans who, however, have two instruments in orbit.

Despite being cold and desert-like, Mars is a planet full of promises.

P. Caraveo, *Space Ecology*, https://doi.org/10.1007/978-3-031-78344-9_6

Fig. 1 Overview of all the missions of exploration of the solar system up to 2012 (Credit National Geographic) listing space missions, both successful and failed ones, for different agencies. Success is indicated by a solid line, and failure by a dashed line. The agencies listed are NASA (orange), U.S.S.R./Russia (red), European Space Agency (green), Japan (purple), China (yellow), and India (blue)

Since the first grainy images of the NASA Mariner IV mission in 1965, the mapping of the Martian terrain from orbit shows without a shadow of doubt the presence of a dry river system demonstrating that on Mars, in a remote past, water flowed. Therefore, the planet must have had an atmosphere and must have been a more welcoming place, warmer and certainly more humid than the current desert. Could it have possibly hosted some form of life?

The images of the dry rivers, along with the persuasive ability of Carl Sagan, had convinced NASA to send probes equipped with an analysis laboratory to search for organic material in soil samples collected by a mechanical arm. Thus the two Viking probes, each composed of an orbiter and a lander, after performing a perfect landing, started to work without however reaching clear results. Sometimes a signal coming from organic molecules seemed to be present, other times there was nothing. In fact, the data were contradictory and they were never understood until, in 2008, the Phoenix lander, which landed in the polar area, solved the enigma revealing the presence of perchlorate, a compound that was not expected to be found, capable of contaminating the results.

After the success of the Vikings, Martian exploration did slow down, marred by a series of fiascos. It returned to the spotlight in 1996, thanks to the Mars Global Surveyor probe and the Pathfinder lander that brought the small rover Sejourner, the first to travel a few meters on the surface of the planet demonstrating the feasibility, and the great potential, of mobile exploration on the red planet.

In parallel, still in 1996, the scientific world was shaken by news related to the analysis of ALH84001, the most famous of the Martian meteorites. It is a 2 kg stone collected in Antarctica on the Allan Hills in 1984 but jumped to the headlines on August 7, 1996 when then-President

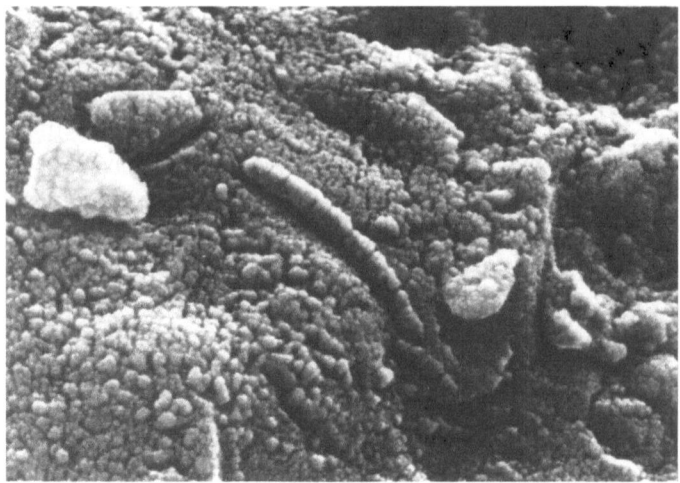

Fig. 2 The electron microscope image of the Martian meteorite ALH84001. (Credit NASA)

Bill Clinton, during a crowded press conference, said "Today this piece of rock speaks to us, and it speaks to us through distances of millions of miles and times of millions of years: it speaks to us of the possibility of life". Life on Mars? A truly extraordinary statement that brought to mind Carl Sagan who used to say *extraordinary claims require extraordinary evidence.* But, where were the "extraordinary proofs"? The figure shows a detail of the photo that went on the front pages of all the newspapers of the world (Fig. 2).

The structure in the center, similar to a segmented worm, had been interpreted as a form of Martian life somehow "fossilized". Given the profound implications, a dispute immediately began among experts from various disciplines to determine whether the structure was biogenic, i.e. the result of activity linked to elementary forms of life, or if it was simply of mineral origin. Clearly, the evidence was not extraordinary enough, but it was a fascinating hypothesis. The mere fact of talking about it brought astrobiology to the attention of the general public, under the media spotlight, and produced the effect that the NASA administrator had hoped for: an increase in the budget for planetary exploration. If we have had a flourishing of Martian missions in the new millennium, we must thank ALH84001, even if, as time went by, the evidence of fossil life did become increasingly thin.

Even without a fossil worm, however, the search for life on Mars continued following completely different paths, for example by hunting for biomarkers, i.e., gases produced by biological activity. Thus began the epic of measuring methane in the thin Martian atmosphere. Let's remember that methane is a particularly interesting gas because on Earth it is mostly produced by living beings (animals and plants) but it can also have a geological origin. The measurement of methane in the atmosphere of Mars is a controversial story with continuous twists and turns. It's there, it's not there, it's there, but only sporadically, are the messages that have come from the probes that *sniff* Mars both from orbit and on the ground. Mars Express (orbiting probe of the European Space Agency that, since December 2003, monitors the red planet) revealed it, first, years ago; then again recently, but this time in conjunction with NASA's rover Curiosity, operational since 2012. Indeed, Curiosity, with its very fine sense of smell, has measured periodic variations in the amount of methane present in the atmosphere, giving rise to the hypothesis that it was a seasonal event. The rise in temperature during the Martian spring awakens, perhaps, the methane producers, or perhaps, releases pockets of gas of geological nature. The most consistent detections made by the Curiosity mission hover around twenty parts per billion (ppb). Trifles compared to the over 1800 ppb of methane in the Earth's atmosphere, but on Mars there are neither cows nor rice fields, which are among the most important producers of terrestrial methane. The data are confirmed by the Trace Gas Orbiter (TGO) also of the ESA which has been in Martian orbit since 2016.

Although infinitely less abundant than on Earth, the presence of Methane in the thin Martian atmosphere is a clue that cannot be overlooked. What produces it?

In addition to the volatile methane, the Martian subsoil hides other secrets. The most intriguing one was announced in 2018 by a team of Italian scientists, led by Roberto Orosei of the National Institute of Astrophysics (INAF), using an instrument conceived and built in Italy that operates on board the European probe Mars Express.

This is MARSIS (Mars Advanced Radar for Subsurface and Ionosphere Sounding), a radar that uses very long waves to be able to penetrate up to a few kilometers into the Martian crust. The same technology is used in Antarctica and Greenland or in the Canadian Arctic to search for subglacial lakes, exploiting the different behavior of water and ice w.r.t radar signals. While ice, especially if it is very cold, is transparent to radar waves of this type, water is not.

When the radar waves hit the Martian polar cap, some are reflected and some penetrate the stratified ice. If, at a certain depth, they encounter something less transparent, the waves are reflected again and the instrument in orbit records a second return signal. From the difference in arrival times between the first and the second signal, the depth of the obstacle can be calculated.

By repeatedly observing a region of the Planum Australe over the course of several years, MARSIS has seen a strong return signal from an area of about 20 km in diameter at depth of 1.5 km that appears to be a subglacial salt lake. Perhaps this is where the ancient Mars water is hiding. The very same water that, at the beginning of the planet's life, flowed on the surface and filled the lake basins and the riverbeds that we see perfectly dry in the photos of the probes in orbit. It had always been suspected that part of the water had hidden underground and now we see the proof.

Mars never goes out of fashion: in 2021 it filled again the scientific chronicles. Three probes, launched in July 2020, arrived at destination in February 2021. The martian bonanza started with the insertion into orbit of the first probe from the United Arab Emirates called Al Amal (Hope), then came Tienwen-1 (questions to the sky), the first Chinese mission that began to orbit while preparing for the landing maneuver of the rover Zhurong (god of fire), successfully carried out in May. NASA, on the other hand, chose a different strategy by proceeding without hesitation to the landing of the Perseverance probe. The spectacular maneuver was captured, for the first time, in high resolution by the onboard cameras. Then, in April 2021, came the maiden flight of the small helicopter Ingenuity which passed the test so brilliantly that it was promoted from a technological test to a component of the mission (Fig. 3). Its view from above guided Perseverance in the search for interesting rocks to be examined and, if necessary, collected with the onboard drill. Its extraordinary mission ended

Fig. 3 The first image of Ingenuity ready to take flight. (Credit NASA)

after over 70 flights in January 2024, when one of the blades of the rotors was broken during landing on a dune too steep.

The samples collected by Perseverance are sealed in special containers that will wait for another probe to retrieve them and start the long journey home, where they will arrive on board a third probe. A complicated scheme, which will require quite a bit of time but the study of Mars is well summarized in the message hidden in the multicolored parachute of Perseverance. The solution to the riddle is a beautiful quote from Theodore Roosevelt: **Dare mighty things**.

Unfortunately, the Europeans, who planned to be at the forefront with the ExoMars mission, a collaboration between the European Space Agency and the Russian one (Roscosmos), are missing on the red planet. ESA was to provide the rover with an innovative drill capable of reaching a depth of 2 m while Roscosmos was responsible for the launch with a Proton as well as for the landing system, capable of depositing the rover on the planet's surface. The mission, originally planned in 2018, was not deemed ready for departure and was postponed to the next launch window in July 2020, the centenary year of Rosalind Franklin's birth, to whom the rover was dedicated. The pandemic, along with problems with the parachutes, prevented the launch in July 2020 and everything was rescheduled for September 2022. When both the Russian lander Kazachok and the European rover Rosalind Franklin were about to be shipped to the Baikonur cosmodrome, the consequences of the tragic invasion of Ukraine caused a rapid deterioration of international relations prompting ESA's decision to suspend the launch. And with Mars, you don't mess around. Launches can take place during windows of a few weeks that repeat every 26 months, when the reciprocal motions of Earth and Mars bring them to the minimum distance. Rosalind Franklin will have to wait until 2028.

The Unknowns of Human Exploration of Mars

To prepare for human exploration of Mars, we need to thoroughly understand the conditions that astronauts will face. We know that the environment is not hospitable, the temperature rarely exceeds 0° and vast portions of the planet can be affected by terrifying sandstorms that can last for weeks, rendering solar panels useless for energy production. Moreover, Mars has no magnetic field to shield humans from cosmic rays and its atmosphere is too thin to provide protection from meteorites, the number of which turns out to be higher than expected. This is a surprising discovery made by the seismograph of NASA's InSight instrument, which, while studying Martian seismology, revealed many extremely short episodes due to the impact of meteorites the size of a basketball. On Earth, these celestial visitors burn in the atmosphere but this does not happen on Mars where Insight has estimated that each year between 280 and 360 meteoroids the size of basketballs arrive, digging craters over 8 m in diameter. The rate of almost one a day is about five times higher than the number estimated from images collected by satellites in Martian orbit and from extrapolation of the number of impacts on the Moon. The abundance of meteorites is linked to Mars mass (and thus gravity), greater than that the Moon, and to Mars position, physically closer to the asteroid belt. This discovery, in addition to its astronomical interest, highlights a new risk for future Mars explorers.

All things considered, however, the greatest risk remains that related to the dose of cosmic radiation that explorers will absorb during the long interplanetary journey when it is very difficult to imagine effective protections. The problem, although present on Mars, could be mitigated by building settlements underground, exploiting lava tunnels, naturally shielded by the overlying material that absorbs much of the radiation. A solution that does not seem to please Elon Musk who has repeatedly stated that he wants to invest his wealth to build a thriving Martian colony. Thinking big, he wants to build a vast settlement where settlers would live and grow their food in transparent domes. The billionaire, who does not seem particularly interested in searching for signs of present or past life on our neighboring planet, has instead stated that he is firmly determined to go and die on the red planet.

To warm up Mars, the only hope might come from the emission of greenhouse gases into the thin atmosphere. Hence the idea that seems to have seduced him: nuke Mars, i.e. bomb the polar ice caps of Mars, where there are deposits of water ice and carbon dioxide, to release the greenhouse gas (responsible for the warming of our planet) that would have beneficial effects on the Martian climate.

Let's say that Musk is certainly not the first to wonder if it would be possible terraforming Mars but those less hasty than him have started by asking themselves whether there is enough carbon dioxide, and, in general, enough gas, to hope to increase the atmospheric pressure and temperature on Mars. The data from the MAVEN probes (for Mars Atmosphere and Volatile EvolutioN which analyzes how Mars loses its atmosphere) and the images from Mars Express, along with those collected by the Mars Reconnaissance Orbiter and Mars Odyssey, were used to estimate the amount of carbon dioxide in the polar ice caps. Even taking into account the carbon dioxide trapped in the rocks, the Martian census does not bode well and those who have seriously done some calculations have concluded

there is not enough CO_2 remaining on Mars to provide signifi-cant greenhouse warming were the gas to be put into the atmosphere; in addition, most of the CO_2 gas in these reser-

voirs is not accessible and thus cannot be readily mobilized. As a result, we conclude that terraforming Mars is not possible using present-day technology.

So it is not advisable to launch nuclear bombs on the Martian poles. An exercise that, apart from being useless, could prove harmful since the explosions could release so much dust as to cause a nuclear winter, further cooling the already freezing Mars.

Planning to colonize another planet with nuclear explosions demonstrates a total absence of any ecological consciousness.

Planetary Defense

After talking about the impact that the use of space has on the ecology of the circumterrestrial environment, as well as on that of the celestial bodies that have been explored so far, we broaden our vision to include interactions between the terrestrial environment and the other bodies of the solar system, touching on a topic that is destined to assume an increasingly important role in the years to come: planetary defense. It is a term that evokes almost catastrophic scenarios where we must defend our planet from the impact of killer asteroids. While this is undoubtedly a correct interpretation, it is not the only one. Planetary defense also has a biological dimension that sees us doubly protagonists, since we must avoid contaminating celestial bodies we visit with terrestrial organisms and, conversely, we must ensure that samples of extraterrestrial material, laboriously collected and brought back to earth, do not contaminate our world.

The biological dimension of planetary defense emerged very early, well before the moon landing of Apollo 11 and the arrival of lunar samples. In fact, the subject is regulated

P. Caraveo, *Space Ecology*, https://doi.org/10.1007/978-3-031-78344-9_7

by Article IX of the treaty for the use of outer space (see Appendix) which came into force on October 10, 1967. Precisely in view of missions to other bodies of the solar system, the article states that exploration activities must avoid both harmful contamination of celestial bodies and harmful consequences related to the arrival of extraterrestrial matter on our planet.

In this regard, the Committee on Space Research (COSPAR), born in 1958 to allow exchange and dialogue between all the space powers, has always been the focal point of international discussions on planetary defense and, acting with the consensus of all member states, has developed an approach based on the classification of space missions into 5 different categories depending on the type of mission (close passage, orbiting mission, ground mission) and the potential interest for the presence of life in the celestial body under study. Type I missions are those that study objects that are not believed to be able to host life, type II missions are directed at potentially interesting bodies for which contamination is not feared (for example, a brief flyby mission). Category III is assigned to missions aimed at orbiting interesting bodies for which extreme cleanliness must be anticipated, since the probe could end up hitting the surface.

Everything that lands on the surface of a celestial body falls into category IV which requires extreme cleanliness complete with sterilization procedures to minimize the possibility of contaminating planetary environments with terrestrial material. Sample return missions that bring back extraterrestrial samples are category V, implying that the material be considered potentially dangerous, and stored in environments with biological safety level 4, the most rigorous possible (the one used to study Ebola).

The classification of a mission is not cast in stone, since the data collected can lead to a change in classification. Based on the results, a type II mission can become a type III mission as happened to the Galileo probe after the discovery of the promising features of Europa, whose icy surface protects an ocean where some form of life may have developed. To prevent the probe, at the end of its orbital life, from running the risk of crashing into Europa, with the danger of contaminating it, it was decided to end the mission with a plunge into Jupiter. A similar procedure was followed for the Cassini probe, which was made to fall on Saturn to avoid possible contamination of the promising Enceladus or the larger Titan.

On the one hand, it is necessary to sterilize (as much as possible) the spacecraft, which are kept in clean rooms carefully controlled to minimize the risk of direct contamination through bacteria, viruses, fungi or spores that, traveling on the equipment, could colonize biospheres in other parts of the Solar System. On the other hand, it is necessary to carefully plan the management of what comes from space that could contain alien pathogens, potentially dangerous for life on earth.

These two issues, which we now deal with in parallel, have developed at different times. While NASA was concerned about preventing contamination on Earth of everything that, following the Apollo program, came from the Moon, the awareness of the need to avoid contamination of the lunar soil only came later, when scrap metal and various types of terrestrial garbage had already been deposited on the Moon.

The Imperfect Quarantine of the Apollo Missions

In fact, the Apollo program was the first example of planetary defense where the Earth felt the need to defend itself from potential alien pathogens. For years, before Apollo 11, experts wondered about the possibility that the Moon could host some form of life. In the 60s no one could say for sure. Of course, the conditions of the lunar soil without atmosphere, without magnetic field, exposed to extreme thermal excursions and continuously sterilized by the ultraviolet radiation of the Sun did not seem ideal, but scientists were worried enough that the National Academy of Sciences organized a high-level conference in 1964 to discuss Moon-Earth contamination.

In other words, what could have happened if lunar microbes had survived the return journey? The conference participants agreed that the risk, although remote, was not nil and that the consequences could be profound. For this reason, it was necessary to quarantine anything that returned from the Moon even though many suspected that, as necessary as this move was, it could be useless since it seemed really difficult to contain a microscopic threat, especially if it had traveled with the astronauts in the reentry module.

Despite these conclusions, NASA has publicly claimed to be able to protect the planet and has spent tens of millions of dollars on a sophisticated quarantine structure, the Lunar Receiving Laboratory, where the astronauts from the first lunar missions were kept in quarantine for 21 days. It was not a simple exercise because keeping human beings in complete biological isolation is very complicated, and recent studies of the protocols and methodologies used at the time have revealed that, in fact, the isolation was far from perfect. Archive documents show that NASA officials were well aware of the laboratory shortcomings. During tests and inspections, the sterilizing autoclaves broke down or had leaks and caused flooding. In addition, the emergency procedures for the laboratory—to be followed in case of fire or medical problems—implied breaking the isolation.

In the weeks following the return of the Apollo 11 crew, 24 people were exposed to the lunar material from which the infrastructure of the Lunar Receiving Laboratory was supposed to protect them and had to be put in quarantine.

However, despite the errors in managing the quarantine of human beings, the real weak link in the chain of protecting the planet from possible (although very unlikely) lunar pathogens was precisely the triumphant moment of the capsule's splashdown in the Pacific Ocean. We remember the footage with a team of frogmen who went to recover the astronauts. Floats were attached to the capsule, then someone climbed up and knocked on the hatch that had to be opened to throw in the package with the biological containment suits that the astronauts had to wear before exiting. What about lunar pathogens, carried by the two astronauts who had been on the Moon, hidden somewhere in the capsule waiting for their chance to come into contact with the air or with the water through the hatch that necessarily had to be opened to allow the astronauts to exit? While the outside of the capsule had certainly been sterilized by the very high temperatures developed as a result of friction with the atmosphere during reentry, inside there could be anything. The problem had not escaped the attention of NASA's planetary protection experts who, in a 1965 document, wrote that the agency was morally bound to prevent potential contamination, even if this meant modifying the weight, costs, or schedule of the mission. Too bad that any modification to a complex mission like the Apollo project would have caused delays and NASA was in a hurry. Four years later the capsule splashed down, showing that NASA had accepted the risk (very low, indeed) on behalf of the entire planet.

"If lunar organisms capable of reproducing in the Earth's ocean had been present, we would have been toast", said John Rummel, who served two terms as NASA's planetary protection officer. The probability that such organisms existed was very small. But if they had been there and had managed to reproduce, the consequences would have been enormous.

"This ended up being an example of planetary protection security theater," said Jordan Bimm, a science historian at the University of Chicago.

The photos of the smiling astronauts waving to President Nixon from inside their contamination box were a way of reassuring the public since many knew (or suspected) that the efforts of "planetary protection" were inadequate.

But the astronauts had also brought many kilograms of lunar material either in bags, like the one filled by Neil Armstrong just after descending from the LEM, or in sealed containers. In this case, the danger was twofold because, in addition to preventing the lunar material from contaminating the Earth's environment, it was necessary to protect the precious samples from any possible contact with terrestrial material.

The Strange Stories of Lunar Material

When it comes to rocks and dust, the management of lunar samples seems simpler than that of the crews. The more than 380 kg of lunar material brought back from 6 Apollo missions found shelter in the Lunar Sample Building, a building entirely dedicated to them and their preservation, at the Lyndon B. Johnson Space Center in Houston. To avoid terrestrial contamination, the samples are stored in a pure nitrogen atmosphere inside sealed containers, and are handled only with special tools. No one can even remotely imagine touching them. Their analysis has allowed us to understand that the Moon is a piece of Earth, but it has also clarified the chronology of the formation of the crust, which has been dated to 4.4 billion years ago.

About 2/3 of the lunar samples are still stored in their original containers that have never been opened. NASA wants to preserve the lunar heritage in the belief that future tools will allow new research.

However, a small part of the lunar material has taken other paths and fragments of rocks and dust, suitably encapsulated in transparent plastic packages, have been used as prestigious gifts, even if they do not look very conspicuous.

The astronauts of the Apollo missions gave them to the presidents of the states they visited on their international propaganda tours. In total, at the end of the Apollo 11 mission, 192 plaques containing lunar dust were prepared, followed by 135 plaques with material brought home by Apollo 17. The gifts have not always been carefully preserved and about 70% of them have been lost.

If the lunar samples donated abroad are hard to find, even those donated to the 50 states of the Union by President Nixon are not under control. After a few years, many of these samples had been lost, simply because no one had thought to register them. Only meticulous research has allowed to locate almost all of the 50 lunar samples from Apollo 11, but some are still missing.

The small boxes with a pinch of lunar dust had also become a farewell gift for NASA engineers who were retiring.

It's normal that they were put in a drawer and forgotten, until accidental discoveries by the heirs who thought of monetizing the memory. This did not go unnoticed and, when NASA sees these announcements, it asks (and generally obtains) that the material be seized.

Obviously, NASA is not short of lunar samples, but it is a matter of principle. Anyone who has received a gift box is certainly authorized to keep it but not to sell it. A policy aimed at not creating a parallel market for lunar material.

There have been lawsuits, which, at times, have also seen former astronauts (or their heirs) among the defendants where the judges have always sided with NASA. But all rules have exceptions and, when it was NASA that made glaring mistakes, the judges were forced to rule against the agency.

This happened in 2015 when Nancy Lee Carlson, a Chicago lawyer with a passion for the history of space missions, bought at an auction a memento of the Apollo missions: it was a bag with the inscription Lunar Sample Return that obviously had been used to bring back lunar samples. In the auction catalog the object was attributed to the Apollo 17 mission. Once she received the bag (which had cost her just under $1000) the lady thought of sending it to NASA to get a certification of what she had bought. At NASA it was immediately recognized that it was the bag that had been used by Neil Armstrong and of which they had lost track following the theft of material at the Kansas Cosmosphere and Space Center in 2003. Once the stolen goods were recovered, the object had been cataloged as part of the Apollo 17 equipment (which, however, did not have this type of containers).

NASA immediately asked for the return of the historic bag, but the woman responded that she had bought it in a legal auction and therefore it was her property. The case went to court and the judge ruled in favor of the woman: NASA had to return the bag, which went up for auction in 2017 and was sold for 1.8 million dollars. However, NASA had retained the dust that was attached to the fabric (it was thanks to this that they had understood beyond any doubt that it was material from Apollo 11) and did not intend to give it back to Mrs. Carlson, who sued the agency to get back what was hers. In this case too, the judge ruled in favor of the woman who auctioned off the aluminum containers containing 0.2 g of lunar dust. The bids stopped at 504,375 dollars, a nice sum that makes lunar dust one of the most expensive materials on the market.

When Visits Contaminate

So far we have talked about material coming from the Moon, now let's change perspective and consider what happens when we are the visitors. We do not know how accurately the Soviet and American probes, which first crashed and then landed, more or less gently, on the lunar surface in the 60s, were sterilized, but, certainly, one cannot think of sterilizing the astronauts. Giving for granted that, despite all precautions, it is impossible to eliminate the biological contamination associated with the presence of astronauts moving on the lunar soil, it would seem that the Apollo program did not comply with Article IX of the treaty that says to avoid harmful contamination of the surrounding environment. In hindsight, perhaps it would have been better to bring home the garbage (biological and otherwise) rather than leaving a small dump of all the material that was to be disposed of before taking off to rejoin the command module that was waiting for them in lunar orbit. Undoubtedly, at the time, lunar samples were considered far more important than garbage, much to the chagrin of lunar ecology. Certainly, the return to the Moon programs will be more respectful of the environment, even if not everyone seems to understand what is to be considered harmful contamination of the lunar soil. We find a recent example with the Israeli mission Beresheet, which was supposed to land on the Moon in September 2019, but instead miserably crashed. Only after the fact, it was announced that, in addition to memories and cultural information, the probe had on board a certain number of tiny dried tardigrades. The idea of secretly sending a sample of the most resilient animals known to the Moon appears bizarre, to say the least, and constitutes a blatant violation of Article IX of the OST.

What About Mars?

Obviously, all the precautions taken for lunar missions are small compared to what would be necessary for the exploration of Mars where, for sure, 3 billion years ago water flowed and where some form of life could have developed. This is why it is so interesting to study Mars, also to understand where the water, that left unequivocal signs of its presence in the rivers' beds now dry, has gone. The Curiosity and Perseverance probes are precisely studying, respectively, an estuary and the bed of a lake. Neither Curiosity nor Perseverance are equipped to search for life forms: they are rather geologists looking for the signatures of the presence of water in the chemistry of the rocks that are photographed and, when necessary, drilled. They collect and analyze samples of the atmosphere also to understand the origin of the "whiffs" of methane that occasionally appear, often in conjunction with the Martian spring. It is a topic of great interest since methane can be produced by plants and animals or it can have a geological origin. Understanding the origin of methane on Mars would be an important step in clarifying the possible presence of microorganisms.

Even if they do not have mini laboratories to search for the presence of organic material, as the Viking did (without achieving convincing results) at the end of the 70s and, more recently, the Phoenix probe, the rovers have been carefully sterilized. In addition to being built in super-clean and super-controlled environments, the vehicles were packaged and sterilized with procedures that should almost completely eliminate any type of spore. Precautions must be even more stringent in the case of drillers who want to go and collect samples 1 or 2 m deep, as the Rosalink Franklin probe of the European Space Agency should do, when it will finally operate on the red planet. While the solar wind and the UV radiation of the Sun make the surface an inhospitable place for any forms of life, things underground could be different and certainly no one wants to send a driller to Mars to find germs of terrestrial origin. Such precautions are not needed to operate in Torino, at the Thales Alenia Space Mars simulator, the probe twin which has been named after Amalia Finzi, one of the most prominent space scientist in Italy.

However, no robot working on Mars, no matter how "intelligent", will ever be able to perform analyses comparable to those possible in terrestrial laboratories. This is why for quite some time a mission to collect and bring Martian samples to Earth has been in the planning with the goal to search for evidence of some kind of metabolism or microfossils.

Once collected, the samples are to be brought into Martian orbit where they must find a return probe provided by the European Space Agency to trasport them back to Earth. This is a very complex project, known as Mars Sample Return, which, at this moment, is in a pause because the latest cost assessments have sounded an alarm bell and NASA wants to rethink the entire architecture to try to lower them.

Fig. 1 Sealed tube containing samples collected by Perseverance, deposited on December 21, 2022 awaiting the mission to recover it and bring it to Earth. (Credit NASA)

This does not prevent Perseverance from continuing to collect samples of the rocks that seem most interesting and which, sealed in special containers, are then deposited in well-defined places to allow a future mission to recover them and bring them back to Earth (Fig. 1). They would arrive in a sealed capsule that would land in the Utah desert, as was the case for the return of the samples from the Bennu asteroid collected by the Osiris Rex mission.

While the Bennu samples were treated with all precautions to prevent contamination of terrestrial material and were handled with a procedure similar to that of lunar rocks, the approach to the management of Martian samples shoud be different.

Carl Sagan began to think about it in his book "The Cosmic Connection", published in 1973, where he imagined that life existed on Mars. Elaborating on this fanciful scenario he wrote "it is possible that on Mars there are pathogens, organisms which, if transported to the terrestrial environment, might do enormous biological damage—a Martian plague.". As a very experienced planetologist, however, Sagan added "The likelihood that such pathogens exist is probably small, but we cannot take even a small risk with a billion lives".

Certainly, no one can say with certainty that the samples, when and if they arrive, will not contain Martian microorganisms. If this were the case, no one can yet say with certainty that they are not harmful to Earthlings. For this reason, NASA must act as if the samples coming from Mars could generate the next pandemic. Andrea Harrington, the Mars sample curator for NASA, explained that the agency plans to handle Martian samples in a similar way to how the Centers for Disease Control and Prevention handle Ebola: with great care.

This means that once the Mars samples arrive on Earth, they must be stored in a facility called the Sample Receiving Facility, which should meet a standard known as "Biosafety Level 4" or BSL-4, which means it is capable of safely containing the most dangerous pathogens known to science. At the same time, however, the facility must also prevent contamination of the samples: a gigantic clean room that prevents substances present on Earth from contaminating the samples coming from Mars.

Materials from all over the solar system have already arrived on Earth to be studied: lunar rocks and dust from American, Soviet, and Chinese missions; samples from three asteroids collected by Japanese and American probes; and particles from the solar wind and a comet collected by NASA missions. But Mars presents what NASA considers a "significant" risk of contamination.

"We have to treat these samples as if they contain hazardous biological materials," said Nick Benardini, NASA's planetary protection officer.

John Rummel, whom we have already met for the Apollo missions, believes it is right that the space agency takes the risks seriously, even if they are small and seem like science fiction. "There are significant unknowns with respect to the biological potential. A place like Mars is a planet. We don't know how it works".

The biggest technological challenge is that the sample receiving facility must meet two radically different purposes. The Earth must not pollute the sample (so nothing should be able to enter the Receiving Facility) and, conversely, the sample must not come into contact with the Earth (so nothing should come out).

The function of a high containment laboratory is precisely to keep what is inside, inside, but in the case of Martian samples this requirement is joined by the equal and opposite one: what is outside must stay outside, complicating the design of the Mars Receiving Facility for which construction is estimated to take between 8 and 12 years. Meanwhile it will be necessary to develop the procedures for sterilizing and then decontaminating the tools used for handling the samples, which, while remaining isolated, must be analyzed. The temporary halt of the Mars Sample Return mission has allowed more time to clarify ideas and also to think about the best communication strategy.

Public reactions could range from extreme concern about the handling of potentially dangerous samples to difficulty in believing that all this is necessary. After all, more than 100 meteorites have naturally arrived from Mars and they have been collected and analyzed prior of having information about their nature, which was understood only afterwards thanks to the presence of Martian atmosphere trapped in the rocks.

In other words, the feeling about Martian material ranges from Carl Sagan's apocalyptic fears to the flourishing market of Martian meteorites passing through ALH84001 with its pseudo Martian worm.

Towards a Sustainable Space

The launch of Sputnik, humanity's first satellite, in 1957 marked the dawn of a new era.

Decades later, our planet is surrounded by satellites that do extraordinary work in studying climate change, saving human lives and mitigating the consequences of natural disasters, providing global communication and navigation services and helping us answer important scientific questions.

Certainly, rapid technological developments have prompted innovative applications for the benefit of society. Unfortunately, however, negative impacts are emerging, such as light pollution, the danger of collisions, the deposit of toxic gases in our atmosphere up to the risk of accidents caused by free-falling debris.

The explosive growth in the number of objects in orbit implies global risks that cannot be ignored. Accidental collisions between objects in space, the more likely the greater the number of satellites traveling in orbits at the same height, produce enormous clouds of debris that can damage other satellites with an inexorable cascade effect that would

P. Caraveo, *Space Ecology*, https://doi.org/10.1007/978-3-031-78344-9_8

render off limits the most useful orbits around Earth, since they would be no longer safe for spacecraft and people.

As we have seen, satellites already have to perform numerous collision prevention maneuvers to avoid possible impacts with other satellites and with the ever-increasing debris. These maneuvers are costly and hundreds of alarms are already issued every week to avoid collisions.

And this is nothing compared to what is about to happen. Several companies have started launching mega-constellations into low Earth orbit to provide global Internet access. Their services offer great benefits, but could be a source of huge inconvenience if we do not change our approach to the use of space, which is a resource freely available but it is not infinite.

In June 2021 during the G7 leaders summit in Carbis Bay, Cornwall, delegates from Canada, France, Germany, Italy, Japan, United States, United Kingdom and EU committed to act to address the growing danger of space debris as our planet's orbit becomes increasingly crowded. Together they declared

We are committed to the safe and sustainable use of space to support humanity's ambitions now and in the future.

We recognise the growing hazard of space debris and increasing congestion in earth's orbit.

As the orbit of our planet is a fragile and valuable environment that is becoming increasingly crowded, which all nations must act together to safeguard, we agree to strengthen our efforts to ensure the sustainable use of space for the benefit and in the interests of all countries.

We welcome the United Nation's Long Term Sustainability Guidelines and call on others to join us in implementing these guidelines.

We welcome all efforts, public and commercial, in debris removal and on-orbit servicing activities and undertake to encourage further institutional or industrial research and development of these services.

We recognise the importance of developing common standards, best practices and guidelines related to sustainable space operations alongside the need for a collaborative approach for space traffic management and co-ordination.

We call on all nations to work together, through groups like the United Nations Committee on the Peaceful Uses of Outer Space, the International Organization for Standardization and the Inter-Agency Space Debris Coordination Committee, to preserve the space environment for future generations.

Excellent intentions that demonstrate that the problem of orbit congestion has reached political leaders who, however, have not undertaken collective actions. Finding global consensus on initiatives that necessarily should impose limitations on space operators is very difficult. The problem of orbital crowding should be addressed with the same determination that led to the Montreal Protocol for limiting the production of CFCs (Chlorofluorocarbons), gases responsible for the growth of the ozone hole. The evidence of the destruction of the Ozone layer, which protects us from the Sun's UV radiation, was considered a threat to life on the planet and nations agreed to suspend the production of CFCs replacing them with other compounds. Entered into force on January 1, 1989, the protocol represents, according to former UN Secretary Kofi Annan, *perhaps the single most successful international environmental agreement to date.*

While the overcrowding of Earth's orbits is not yet perceived as a global danger, time is running out and the situation continues to worsen. The space industries might start to move. Recognizing the existence of a serious problem, which can put at risk the development of their activities, they should consider voluntary initiatives to promote the sustainability of their business model.

Since there is no universal definition of sustainability in space or how to achieve it, the World Economic Forum, in collaboration with the European Space Agency, the Space Enabled Research Group of the MIT Media Lab and the University of Texas at Austin, has proposed the Space Sustainability Rating (SSR). It is a new system for assessing space sustainability that is quantified on four levels. The SSR is aimed at satellite service operators, launch service providers, and satellite manufacturers who can choose to have their operations evaluated.

The scores are based on factors ranging from data sharing, to the choice of orbit, to measures taken to avoid collisions, to plans for the re-entry of satellites at the end of their mission and to their ability to be detected and identified. Also the characteristics of the launcher will have an impact on the score. By voluntarily participating in the evaluation program, industries can obtain a certification that, in addition to improving the international "reputation", could be used to lower insurance premiums or to facilitate fund rasing. If it were subscribed to at the national level by many states, the SSR could become the space equivalent of the ethical certification that accompanies every diamond according to the Kimberley Process Certification Scheme, established in 2002 in Interlaken, to prevent the marketing of "bloody diamonds". The Certification Schene was signed by 37 States, both producers and importers of precious stones.

Even though, at the moment, it is still on a voluntary basis for individual industries, the SSR is an excellent starting point and I hope it can trigger a phenomenon of virtuous emulation throughout the space supply chain.

However commendable, goodwill is not always enough, time is running out and the situation of low orbits continues to worsen. There are thresholds beyond which further development becomes unsustainable, leading to environmental degradation and unnecessary risks.

For example, it is possible to define a limit of occupation for any orbit taking into account the density of objects, technological capabilities and the levels of collision risk. When occupation becomes excessive, the use of that orbit is in fact precluded to new operators who can no longer access a common good that has been overexploited. Along with the number of objects, the disturbance to astronomical observations increases, here too there is a threshold beyond which observations of celestial sources become impossible. Moreover, the impact from pollutants deposited in the atmosphere by the increasingly numerous launches and by the even more numerous controlled and uncontrolled reentries must be carefully evaluated.

Space is a complex ecosystem in constant evolution both from the technological point of view and for its increasingly pervasive impact on our society. When studying solutions for the management of space traffic and the cleanup of debris, their wider impacts on the Earth's atmosphere, on astronomical observations and on the ability of other users to access space must be considered.

While continuing to develop specific technological solutions, a global approach to space sustainability requires shared responsibility, the collaboration of all the actors involved and the involvement of the various parties in the decision-making processes. This requires regulatory and policy measures and incentives at the national and international level, along with courageous leadership and international diplomacy.

What Can We Do?

The first rule that all space operators should give themselves is to **avoid producing space debris**. For everything that must operate in low orbit, this means, first and foremost, designing missions in such a way that they can, at the end of their operational life, lower their orbit to be destroyed by the friction of the atmosphere, while for geostationary missions, it is necessary to move to graveyard orbits where they cannot cause annoyance to operational satellites. To be able to perform these maneuvers safely, satellites must have some kind of propulsion system and residual fuel. These are measures that imply an increase in costs to be budgeted for each satellite and not everyone is enthusiastic about having to apply them. The major space agencies have set standards in this regard, but these are rules accepted on a voluntary basis, everyone is warmly invited to follow them but there is no obligation.

The European Commission would like to adopt a shared space law also aimed at limiting space pollution thanks to a coordinated approach between member states, in order to identify rules to reduce debris, avoid collisions, improve the orbital management of operational satellites and make the entire space supply chain more sustainable.

As we have seen, space debris are not just dead satellites, there is a vast array of fragments of satellites exploded as a result of malfunctions or deliberately blown up during experiments of star war. These are known as DA-ASAT (for Direct Ascent Anti-Satellite) during which a missile launched from earth destroys a target satellite, producing a cloud of debris that continue to move following the orbit of the satellite from which they originate. Over the years, these demonstrative actions have been conducted by China (in

2007), the USA (in 2008), India (in 2019) and Russia in November 2021.

Indeed, to avoid the proliferation of debris that would make low orbits unsafe, hampering the development of space activities. On April 18, 2022, US Vice President Kamala Harris, speaking at the Space Force base at Vandenberg, said that the United States had decided to suspend anti-satellite tests.

Even if it is a unilateral decision, the vice president hoped that others will decide to follow the example. The announcement is a statement of principle since the last American DA-SAT test is from 2008, but it certainly reveals the concern of the US administration about the overcrowding of earth orbits. This does not mean that the Americans have decided not to use space for military purposes. Anti-satellite tests can be conducted using virtual targets that do not produce dangerous debris and, perhaps, can go unnoticed.

In any case, avoiding producing debris is a step in the right direction.

For satellites launched from American territory on low orbit, i.e. with a height less than 2000 km, the Federal Communications Commission (FCC), before issuing the launch authorization, requires a commitment to deorbit non-operational satellites as soon as possible, at most within 5 years. This rule came into effect in September 2022, amending the previous rule, which required that re-entry occur within 25 years. Industries and operators had 2 years to comply with the new regulation, which was designed precisely to avoid crowding orbits with space debris. Reducing the orbital stay after the end of the mission is a first step to not worsen the congestion of traffic in the most requested and busiest low orbits.

To keep space safe and sustainable, on February 12, 2024 SpaceX announced that it would deorbit 100 first-generation Starlink satellites still operational, but, perhaps, obsolete. Considering that the company puts into orbit over 100 second-generation Starlink satellites each month, the gesture has little more than symbolic significance, just to demonstrate that SpaceX has full control of its satellites, something that is not always true for all other operators. As we have seen while discussing February reentering objects, about 4 Starlink satellites do reenter every days since the constellation must continously be updated.

In the case of geostationary orbits, the FCC asks that the satellite that has finished its operational life be transferred to a graveyard orbit 300 km above the geostationary one, where it will not bother the new occupants. This means that whoever operates the satellite must be able to move it 300 km, a maneuver that implies the consumption of a certain amount of fuel. Since there are still no service stations in orbit, the fuel cannot be used up to the last drop for operations, because it is necessary to safeguard that needed for the final maneuver. If the calculations are wrong, the satellite runs out of fuel and stops before reaching the graveyard orbit. This is what happened to the EchoStar-7 satellite, launched in 2002, by the Dish company, which stopped just over halfway. Given that in the documents filed with the FCC the company had agreed that it would stop operating the satellite in May 2022, and that it would move it about 300 km above its operational position, the FCC noted that Dish had violated the Communications Act and fined the company with a penalty of $150,000.

This is the first time this has happened and it is a sign of growing concern about the crowding of orbital space, a reality that has significant implications for the management of space debris. It is an important gesture from the point of view of space law but little more than symbolic from an economic point of view.

In fact, the value of the penalty was negotiated with the company, which, in order to avoid arguing with the powerful FCC, accepted its share of responsibility, reiterating that at the time of the satellite's launch (in 2002) the current rules were not in force and that the poorly positioned satellite posed no danger.

"As satellite operations become more prevalent and the space economy accelerates, we must be certain that operators comply with their commitments" stated in a note Loyaan Egal, head of the FCC's control office. "This is a breakthrough settlement, making very clear the FCC has strong enforcement authority and capability to enforce its vitally important space debris rules." It's a pity that this careful severity is applied only to the limited occupancy GEO orbit, the most commercially attractive, where about 550 satellites operate. For low orbits, say between 500 and 2000 km in height, the numbers are much more significant, but the rules less strict. Moreover, so far we have talked about the FCC which is a federal American body. While it is true that the USA is the nation that launches the most satellites, it is also true that the space landscape is varied and everyone applies the rules they deem most suitable to their needs. Rules that, in any case, cannot be retroactive and are of no help in solving the problem of old debris that may have been in orbit for decades. Even though there is no global law on limiting space debris, providing mechanisms to ensure the safe **de-orbiting** of satellites that have finished their operational life is a measure in the interest of all operators. Space agencies are very careful not to increase space garbage, but it is not always an easy task. NASA has entrusted SpaceX with the construction of a U.S. Deorbit Vehicle, a space tug capable of managing the de-orbiting of the International Space Station when its operational life ends, in 2030. The USDV must be able to drag the structure, which is as large as a football field, from its working altitude of about 400 km to less than 100 km to have it destroyed by the atmosphere, taking care that any remains fall into the Nemo point in the middle of the Pacific Ocean.

Another Approach to Sustainability: Orbit Maintenance

The smallest (and insignificant) failure of a satellite can turn a jewel of technology into unusable junk. For this reason, at the end of the 70s, with the increase of space instrumentation, people began to think about the possibility of doing maintenance in orbit. In fact, this was one of the reasons for the existence of the Space Shuttle, which, using its mechanical arm, could capture satellites that needed maintenance. Then it would be up to the astronauts to go out and replace parts that had stopped working. For this reason, NASA satellites in orbits reachable by the Shuttle were designed with relatively easy-to-open doors and handles to provide a grip for the astronauts. NASA also relied on the Shuttle's cargo bay to transport large loads into orbit. Once the time for release came, the mechanical arm moved the satellite from its parking position keeping it in clear view of the astronauts who could check that all the parts that were supposed to move were in place before releasing it. In fact, human control (and subsequent intervention) saved the Compton Gamma-Ray Observatory, one of NASA's 4 major observatories, launched in 1991. The large platform, which housed 4 instruments for detecting X-rays and gamma rays, was NASA's largest satellite with a weight of almost 16 tons. The observatory had no moving parts except for the solar panels and the antenna for communications with the control center. The astronauts saw that the antenna had remained stuck in the folded position and were forced to make an unplanned exit to convince the antenna to move with a bit of brute force.

The Hubble Space Telescope, another major NASA Observatory put into orbit by the Shuttle Discovery in 1990, was reached and repaired 5 times in the period from 1993 to 2009. However, the era of in-orbit repair never took off because it was a dangerous, difficult and expensive procedure that required the use of the Shuttle. In other words, it brought together the famous 3 D's of Danger, Difficulty and Dollars. Let's not forget that during the last maintenance mission in 2009, when NASA unfortunately had to deal with the loss of the Columbia shuttle on re-entry into the atmosphere and realized the vulnerability of the thermal insulation tiles, it was necessary to keep a second Shuttle on the launch pad ready to take off to go and rescue the astronauts in case something went wrong. The photo of the two Shuttles (Atlantis for the Hubble repair and Endeavour ready, if needed) is a piece of history (Fig. 1).

The closure of the Shuttle program in 2011 ended the chapter on human repair but it was clear to everyone that sometimes the problem could be solved even by an automatic probe capable of performing some essential task, such as helping to deorbit useless satellites or refueling those that had an empty tank but were still perfectly functioning.

After all, no one thinks of throwing away a car when the gas runs out. Just fill up and the car is ready for a new use. This procedure so trivial on Earth (provided there are gas stations) becomes a dream in space. When the propellant tank, necessary to point the satellite in a certain direction or to vary the orbit, is empty, the mission is over. It is natural, therefore, to think of building refueling stations capable of

Fig. 1 The Atlantis shuttle (STS 125 in the foreground) with Endeavour in the background. It's one of the very rare occasions when two Shuttles were ready for launch, the previous one occurred in July 2001. (Credit NASA)

reaching the satellite that has run out of fuel, hooking it up and refueling it, and then letting it go to continue its work. It is clear that having such technology would extend the life of satellites, with obvious economic and ecological benefits. In fact, this would limit the need for new launches, with positive effects on the increasingly pressing problem of excessive occupation of orbits and on air pollution linked to the exhaust gases of launchers and the vaporization of metals during re-entries.

Once again, the issue of refueling had been addressed within the Shuttle program specifically for the Compton Gamma Ray Observatory. Anticipating that the mission would end with the exhaustion of the nearly two tons of hydrazine loaded at departure to allow the observatory to be oriented and maneuvered, NASA had contemplated the possibility of refueling. As early as 1984, contracts had been given to develop the mechanism that would allow the necessary mechanical coupling to carry out the refueling, with particular attention to safety since hydrazine is a very toxic and volatile propellant. In fact, the Challenger disaster, which exploded on takeoff in 1986, put an end to this project deemed too risky.

To be competitive, satellite refueling must be significantly less expensive than a new launch, so the service station must be designed to meet the needs of different customers, who, however, must have tanks equipped with standard docking/refueling valves. It is useless to maneuver the orbiting station to reach a satellite and then discover that mechanical incompatibility hampers docking and subsequent refueling. Even in space, the devil is in the details and the watchword of a potential orbiting refueling station is standardization. For this reason, Orbit Fab began its project, aimed at building a service station in orbit, starting from the tank-user interface. It is called RAFTI (Rapid Attachable Fluid Transfer Interface) and consists of two parts: the structure to allow the mechanical coupling between the refueler and the refueled, which must be firmly attached to each other, and a service valve to be used for filling the tank both on the ground and in orbit.

The RAFTI interface can transfer 1 liter per minute and has been tested on the ISS where it was used to transfer water, but it can be used for a wide range of propellants provided that the tank to be filled is equipped with the right interface. To this end, Orbit Fab is making agreements with operators interested in using the interface for their satellites that could then be refueled (Fig. 2).

The technology developed by Orbit Fab has attracted the attention (and investments) of giants like Lockheed Martin and Northrop Grumman who are interested in acquiring the ability to refuel instruments (civil and military) in orbit. It would be a marked improvement in orbital logistics and

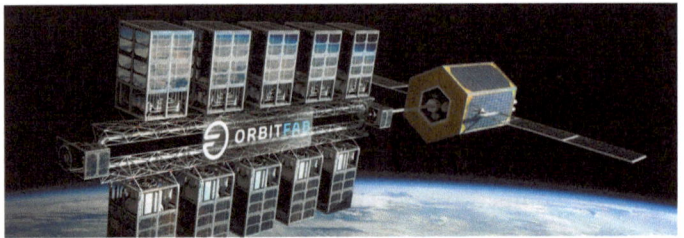

Fig. 2 This is how Orbit Fab imagines the rendezvous between the fuel depot and a satellite to be refueled. (Credit Orbit Fab)

would allow the active life of satellites to be extended, with particular attention to those in geostationary orbit, the most profitable from a commercial point of view. It would be a real shame to have to de-orbit them (freeing up space for other satellites) just because they have run out of fuel. Orbit Fab plans to keep the refueling station about a hundred km from geostationary orbit, and approach the satellite that wants to refuel when the customer requests it. The idea is to reach geostationary orbit, coming from the Moon, in order to take advantage of an Orbital Transit Vehicle from Spaceflight Inc that will be launched by SpaceX to bring the Intuitive Machines Lander to the Moon. A creative solution developed from GeoJump (https://geojump. space/) to exploit lunar traffic in order to reach geostationary orbit.

SpaceX also plans to use a service station as part of the Lunar Lander program that will land NASA astronauts from the Artemis III mission in mid 2027.

Thanks to several Starship launches, SpaceX's plan is to bring the service station into orbit and fill its tank. Once this preliminary step is completed, the lunar lander will be launched which, after refueling, will set off for the Moon where it will wait the arrival of the Orion capsule with the crew.

Space That Offers Models of Sustainability

In addition to showing obvious problems of overuse (and lack of rules) of circumterrestrial space, the conquest of new space frontiers also offers virtuous examples of circular economy that could inspire new approaches to resource management on our planet.

A space colony, whether built in orbit, on the Moon or Mars, must be as sustainable as possible. It is not a choice, but rather a necessity. In space (and on planets) resources, even the most essential ones, are available in limited quantities and must be used to the best of their ability by recycling everything that can be recycled and minimizing waste.

In this perspective, the most precious (and certainly one of the heaviest to transport) is water which, once used, must be recycled starting from the humidity that astronauts produce by breathing and sweating. Similarly to what happens in the driest regions of the earth, where water is extracted from the air, on the ISS the ambient humidity is collected and condensed. But the main source of recycling is the occupants' urine which is filtered and returned to being drinking water. "Today's coffee will be tomorrow's coffee" said an astronaut in connection with a disgusted President Trump. However, no process is 100% efficient and something is lost. In addition, some of the water must also be used to produce oxygen, another fundamental resource to allow life in orbit. For this reason, periodically, cargo arrives with water supplies, in addition to fresh food and other material for the maintenance of the station and all the tools on board. And the ISS is just 400 km from Earth. Water supply on the Moon or on Mars, two celestial bodies on which we want to build permanent settlements, is even more difficult and the survival of future colonies will depend critically on the ability to find it on site, on the surface or in depth. It will not be easy, but not impossible either. The water is there, most likely in the form of ice, both on the Moon and on Mars, but one needs to know where to go to look for it. Hence the need for accurate prospecting of the most interesting areas. In parallel, studies are being conducted on how to use local material for the construction of lunar or Martian dwellings which must be pressurized structures where the astronauts can live without the bulky suits necessary for the external activities. Moreover, both on the Moon and on Mars a barrier must be built to protect the structures from meteorite impacts (not incinerated by interaction with the atmosphere, as happens on Earth) and from cosmic rays (not deflected by the magnetic field, as happens, fortunately, on Earth). Lunar regolith or Martian dust, perhaps compacted with robots equipped with powerful lasers, could be used to cover and protect light structures.

Obviously, cosmic houses must be autonomous from the point of view of energy by transforming the Sun's radiation into electrical energy through solar panels, which must be sized according to energy demands but will never be small. Think of the immense wings of the International Space Station that must also have batteries for the storage of energy to be used when the station passes in the shadow of the Earth. The problem of storage, which is very familiar to us on Earth when we talk about renewable energies, which are inherently intermittent, will be very important for lunar colonies where day and night last two weeks. The situation will be less dramatic on Mars which has a rotation period only half an hour longer than that of Earth.

Once the problem of water, air, and energy is solved, we need to worry about food, which must be produced locally. To grow plants, pressurized environments with water and light are needed (if you want to realize hydroponic crops), so greenhouses, perhaps housed in inflatable structures, lighter and easier to transport are being developed, keeping in mind that they will always need to be protected and shielded from the surrounding environment.

These examples make us understand how space offers virtuous examples of circular economy because, in the absence of raw materials, waste becomes a resource.

Building space colonies means accepting the challenges posed by an extreme and hostile environment by adapting technologies already perfected on Earth, or inventing new ones that, perhaps, will then be used on our planet to make our lifestyle more sustainable.

It's Time to Act

The launch of Sputnik shocked the American public, convinced of having a vast technological superiority over the Soviet Union. Suddenly the oceans were no longer enough to protect them because the harmless Sputnik could be seen as a threat.

While President Eisenhower was preparing to face the launch of Sputnik from the White House, Senate Majority Leader Lyndon B. Johnson contemplated the Congress's response that would lead him to declare

in the eyes of the world, first in space means first. Period. Second in space is second in everything

This belief led him to be an enthusiastic supporter of the *space race*, of the creation of NASA, and of the American space program.

While space continues to have great political strategic value, today we have understood that the future of Earth depends on space. The sustainable development goals, which are the cornerstone of the United Nations vision, could not be achieved without the aid of satellite services.

However, to continue to benefit from space services we must avoid polluting and overcrowding the Earth's orbits.

In his 1997 book Billions and Billions, Carl Sagan wrote "our Technological civilization poses a real danger to itself".

The only way not to run this risk is to learn to respect the common goods that are so indispensable to our well-being and to our lifestyle.

By learning to work in space, we have brought enormous benefits to Earth, providing technologies that enrich our societies, connect people in ways previously unimaginable, and give us a perspective and a deep understanding of our planet.

With the expansion of humanity's space activities, the challenges also increase. The proliferation of satellite constellations offers a great example of the tension between innovation and its consequences, requiring a space policy and a governance system adaptable to the evolution of space activities and their unforeseen impacts. We need to apply to space the lessons we have learned on Earth.

After making many mistakes, we have understood that we have only to gain by not polluting the air and water, by defending nature, by accepting limitations when we use Earth's *global commons*, now is the time to expand our horizons to include the space that surrounds us.

To ensure a balance between progress and conservation, space sustainability strategies must ensure both technological development and the longevity of space "ecosystems".

We know what will happen if we continue on the current path, but we also know exactly what we need to do to change this fate and ensure humanity's access to space for future generations.

Appendix: UN Treaties, International Agreements and National Laws

The exploration and peaceful use of space and celestial bodies are regulated by five Treaties developed within the United Nations between 1967 and 1979. Among these, the most important is certainly the first, known as the Outer Space Treaty (**OST**), which entered into force on October 10, 1967 and has been ratified by 115 nations (with another 23 nations that have signed it but have not yet completed the ratification process). Since it is the cornerstone of space law, I believe it is worth reporting the full text

Outer Space Treaty

Treaty on Principles Governing the Activities of States in the Exploration and Use of Outer Space, Including the Moon and Other Celestial Bodies

The States Parties to this Treaty,

© The Editor(s) (if applicable) and The Author(s), under exclusive license to Springer Nature Switzerland AG 2025
P. Caraveo, *Space Ecology*, https://doi.org/10.1007/978-3-031-78344-9

Inspired by the great prospects opening up before mankind as a result of mans entry into outer space,

Recognizing the common interest of all mankind in the progress of the exploration and use of outer space for peaceful purposes,

Believing that the exploration and use of outer space should be carried on for the benefit of all peoples irrespective of the degree of their economic or scientific development,

Desiring to contribute to broad international co-operation in the scientific as well as the legal aspects of the exploration and use of outer space for peaceful purposes,

Believing that such co-operation will contribute to the development of mutual understanding and to the strengthening of friendly relations between States and peoples,

Recalling resolution 1962 (XVIII), entitled "Declaration of Legal Principles Governing the Activities of States in the Exploration and Use of Outer Space," which was adopted unanimously by the United Nations General Assembly on 13 December 1963,

Recalling resolution 1884 (XVIII), calling upon States to refrain from placing in orbit around the Earth any objects carrying nuclear weapons or any other kinds of weapons of mass destruction or from installing such weapons on celestial bodies, which was adopted unanimously by the United Nations General Assembly on 17 October 1963,

Taking account of United Nations General Assembly resolution 110 (II) of 3 November 1947, which condemned propaganda designed or likely to provoke or encourage any threat to the peace, breach of the peace or act of aggression, and considering that the aforementioned resolution is applicable to outer space,

Convinced that a Treaty on Principles Governing the Activities of States in the Exploration and Use of Outer

Space, including the Moon and Other Celestial Bodies, will further the Purposes and Principles of the Charter of the United Nations,

Have agreed on the following:

Article I

The exploration and use of outer space, including the moon and other celestial bodies, shall be carried out for the benefit and in the interests of all countries, irrespective of their degree of economic or scientific development, and shall be the province of all mankind.

Outer space, including the moon and other celestial bodies, shall be free for exploration and use by all States without discrimination of any kind, on a basis of equality and in accordance with international law, and there shall be free access to all areas of celestial bodies.

There shall be freedom of scientific investigation in outer space, including the moon and other celestial bodies, and States shall facilitate and encourage international cooperation in such investigation.

Article II

Outer space, including the moon and other celestial bodies, is not subject to national appropriation by claim of sovereignty, by means of use or occupation, or by any other means.

Article III

States Parties to the Treaty shall carry on activities in the exploration and use of outer space, including the moon and other celestial bodies, in accordance with international law, including the Charter of the United Nations, in the interest of maintaining international peace and security and promoting international co-operation and understanding.

Article IV

States Parties to the Treaty undertake not to place in orbit around the Earth any objects carrying nuclear weapons or any other kinds of weapons of mass destruction, install such weapons on celestial bodies, or station such weapons in outer space in any other manner.

The Moon and other celestial bodies shall be used by all States Parties to the Treaty exclusively for peaceful purposes. The establishment of military bases, installations and fortifications, the testing of any type of weapons and the conduct of military maneuvers on celestial bodies shall be forbidden. The use of military personnel for scientific research or for any other peaceful purposes shall not be prohibited. The use of any equipment or facility necessary for peaceful exploration of the Moon and other celestial bodies shall also not be prohibited.

Article V

States Parties to the Treaty shall regard astronauts as envoys of mankind in outer space and shall render to them all possible assistance in the event of accident, distress, or emergency landing on the territory of another State Party or on the high seas. When astronauts make such a landing, they shall be safely and promptly returned to the State of registry of their space vehicle.

In carrying on activities in outer space and on celestial bodies, the astronauts of one State Party shall render all possible assistance to the astronauts of other States Parties.

States Parties to the Treaty shall immediately inform the other States Parties to the Treaty or the Secretary-General of the United Nations of any phenomena they discover in outer space, including the Moon and other celestial bodies, which could constitute a danger to the life or health of astronauts.

Article VI

States Parties to the Treaty shall bear international responsibility for national activities in outer space, including the Moon and other celestial bodies, whether such activities are carried on by governmental agencies or by non-governmental entities, and for assuring that national activities are carried out in conformity with the provisions set forth in the present Treaty. The activities of non-governmental entities in outer space, including the Moon and other celestial bodies, shall require authorization and continuing supervision by the appropriate State Party to the Treaty. When activities are carried on in outer space, including the Moon and other celestial bodies, by an international organization, responsibility for compliance with this Treaty shall be borne both by the international organization and by the States Parties to the Treaty participating in such organization.

Article VII

Each State Party to the Treaty that launches or procures the launching of an object into outer space, including the Moon and other celestial bodies, and each State Party from whose territory or facility an object is launched, is internationally liable for damage to another State Party to the Treaty or to its natural or juridical persons by such object or its component parts on the Earth, in air space or in outer space, including the Moon and other celestial bodies.

Article VIII

A State Party to the Treaty on whose registry an object launched into outer space is carried shall retain jurisdiction and control over such object, and over any personnel thereof, while in outer space or on a celestial body. Ownership of objects launched into outer space, including objects landed or constructed on a celestial body, and of their component parts, is not affected by their presence in outer space or on a celestial body or by their return to the Earth. Such objects or component parts found beyond the limits of the State Party to the Treaty on whose registry they are carried shall be returned to that State Party, which shall, upon request, furnish identifying data prior to their return.

Article IX

In the exploration and use of outer space, including the Moon and other celestial bodies, States Parties to the Treaty shall be guided by the principle of co-operation and mutual assistance and shall conduct all their activities in outer space, including the Moon and other celestial bodies, with due regard to the corresponding interests of all other States Parties to the Treaty. States Parties to the Treaty shall pursue studies of outer space, including the Moon and other celestial bodies, and conduct exploration of them so as to avoid their harmful contamination and also adverse changes in the environment of the Earth resulting from the introduction of extraterrestrial matter and, where necessary, shall adopt appropriate measures for this purpose. If a State Party to the Treaty has reason to believe that an activity or experiment planned by it or its nationals in outer space, including the Moon and other celestial bodies, would cause potentially harmful interference with activities of other States Parties in the peaceful exploration and use of outer space, including the Moon and other celestial bodies, it shall undertake appropriate international consultations before proceeding with any such activity or experiment. A State Party to the Treaty which has reason to believe that an activity or experiment planned by another State Party in outer space, including the Moon and other celestial bodies, would cause potentially harmful interference with activities in the peaceful exploration and use of outer space, including the Moon and other celestial bodies, may request consultation concerning the activity or experiment.

Article X

In order to promote international co-operation in the exploration and use of outer space, including the Moon and other celestial bodies, in conformity with the purposes of this Treaty, the States Parties to the Treaty shall consider on a basis of equality any requests by other States Parties to the Treaty to be afforded an opportunity to observe the flight of space objects launched by those States.

The nature of such an opportunity for observation and the conditions under which it could be afforded shall be determined by agreement between the States concerned.

Article XI

In order to promote international co-operation in the peaceful exploration and use of outer space, States Parties to the Treaty conducting activities in outer space, including the Moon and other celestial bodies, agree to inform the

Secretary-General of the United Nations as well as the public and the international scientific community, to the greatest extent feasible and practicable, of the nature, conduct, locations and results of such activities. On receiving the said information, the Secretary-General of the United Nations should be prepared to disseminate it immediately and effectively.

Article XII

All stations, installations, equipment and space vehicles on the Moon and other celestial bodies shall be open to representatives of other States Parties to the Treaty on a basis of reciprocity. Such representatives shall give reasonable advance notice of a projected visit, in order that appropriate consultations may be held and that maximum precautions may be taken to assure safety and to avoid interference with normal operations in the facility to be visited.

Article XIII

The provisions of this Treaty shall apply to the activities of States Parties to the Treaty in the exploration and use of outer space, including the Moon and other celestial bodies, whether such activities are carried on by a single State Party to the Treaty or jointly with other States, including cases where they are carried on within the framework of international intergovernmental organizations.

Any practical questions arising in connection with activities carried on by international inter-governmental organizations in the exploration and use of outer space, including the Moon and other celestial bodies, shall be resolved by the States Parties to the Treaty either with the appropriate international organization or with one or more States members of that international organization, which are Parties to this Treaty.

Article XIV

1. This Treaty shall be open to all States for signature. Any State which does not sign this Treaty before its entry into force in accordance with paragraph 3 of this article may accede to it at any time.
2. This Treaty shall be subject to ratification by signatory States. Instruments of ratification and instruments of accession shall be deposited with the Governments of the United States of America, the United Kingdom of Great Britain and Northern Ireland and the Union of Soviet Socialist Republics, which are hereby designated the Depositary Governments.
3. This Treaty shall enter into force upon the deposit of instruments of ratification by five Governments including the Governments designated as Depositary Governments under this Treaty.
4. For States whose instruments of ratification or accession are deposited subsequent to the entry into force of this Treaty, it shall enter into force on the date of the deposit of their instruments of ratification or accession.
5. The Depositary Governments shall promptly inform all signatory and acceding States of the date of each signature, the date of deposit of each instrument of ratification of and accession to this Treaty, the date of its entry into force and other notices.
6. This Treaty shall be registered by the Depositary Governments pursuant to Article 102 of the Charter of the United Nations.

Article XV

Any State Party to the Treaty may propose amendments to this Treaty. Amendments shall enter into force for each State Party to the Treaty accepting the amendments upon their acceptance by a majority of the States Parties to the Treaty and thereafter for each remaining State Party to the Treaty on the date of acceptance by it.

Article XVI

Any State Party to the Treaty may give notice of its withdrawal from the Treaty 1 year after its entry into force by written notification to the Depositary Governments. Such withdrawal shall take effect 1 year from the date of receipt of this notification.

Article XVII

This Treaty, of which the English, Russian, French, Spanish and Chinese texts are equally authentic, shall be deposited in the archives of the Depositary Governments. Duly certified copies of this Treaty shall be transmitted by the Depositary Governments to the Governments of the signatory and acceding States.

After the OST

In the years following the entry into force of the OST, agreements and conventions were developed to clarify some of the articles.

The Agreement on the Rescue of Astronauts of 1968 expands Article V, which already provided that all signatory States of the OST would commit to providing assistance in the event of accidents to the astronauts, specifying that the term astronauts should refer to the personnel of a spacecraft. Since the agreement was written well before space tourism could be even imagined, it is unclear whether the agreement also covers space tourists. Moreover, while the agreement says that the state responsible for the launch must bear the costs for the recovery of the spacecraft that has performed the emergency maneuver, it does not specify who should cover the costs of a possible rescue of the crew. Following the spirit of Article V, the agreement is about the rescue of astronauts who have been forced to make an emergency landing, but it does not mention the possibility of in-flight rescue, in case of an accident or malfunction of a spacecraft. Once again, this is a sign of the times and space agencies have adapted to changing conditions without feeling the need to modify the agreement. Docked to the International Space Station there are always a number of capsules capable of bringing all occupants back to Earth in case of a serious incident that requires the crew to abandon the station.

The agreement has been ratified by 98 states, 23 have signed (but not ratified) and three international organizations (The European Space Agency, the European organization for the use of meteorological satellites and the Intersputnik organization) have declared their adherence to the agreement.

The Convention on Liability for Damage Caused by Space Objects of 1972;

The Convention says that States have international responsibility for all space objects that are launched from their territory. This means that, regardless of who launches the space object, if it has been launched from the territory of a certain State, that State is fully responsible for the damages resulting from that space object. The Convention was designed to integrate existing and future national laws that provide for compensation of parties injured by space activities. The Liability Convention applies to States and provides that if an individual is injured by a space object and wishes to obtain compensation, he must ensure that his country makes a claim for compensation against the country that launched the space object that caused the damage. The convention was applied only once when a Russian satellite with a nuclear generator crashed in Canadian territory.

The Convention has been ratified by 98 states, 19 have signed (but not ratified) and four international organizations (The European Space Agency, the European organization for the use of meteorological satellites, the European organization for telecommunications satellites and the Intersputnik organization) have declared to accept the rights and obligations of the Convention.

The Convention on the Registration of Objects Launched into Outer Space of 1975;

The convention says that States must provide the United Nations with details on the orbit of each space object in order to have a Register containing information on all launched objects. The register is maintained by the United Nations Office for the Peaceful Use of Outer Space. Knowing to whom a vehicle belongs Space is of fundamental importance in the event of collisions in orbit or damage on the ground.

The Convention has been ratified by 72 states and four international organizations (The European Space Agency, the European organization for the use of meteorological satellites, the European organization for telecommunications satellites and the Intersputnik organization) have declared to accept the rights and obligations of the Convention.

The Agreement on the Activities of States on the Moon and other celestial bodies of 1979.

The Moon Treaty proposes to establish an "international regime" that applies to the Moon and other celestial bodies of the Solar System, including orbits or other trajectories towards or around them.

The treaty, which includes 21 articles, declares that the Moon must be used for the benefit of all States and all peoples of the international community. It reiterates that lunar resources "are not subject to national appropriation by claim of sovereignty, use or occupation, or by any other means". It hopes that the Moon will not become a source of international conflict, so that resources are used exclusively for peaceful purposes.

It prohibits any military use of celestial bodies, including weapons testing, nuclear weapons in orbit or military bases (Article 3.4).

It provides a legal framework to establish a regime of international cooperation to govern the responsible exploitation of the Moon's natural resources (Article 11.5).

It prohibits altering the environmental balance of celestial bodies and requires States to take measures to prevent accidental contamination of celestial body environments, including Earth (Article 7.1).

It hopes for the orderly and safe use of lunar natural resources with fair sharing by all contracting States of the benefits derived from such resources (Article 11.7).

It reiterates that the placement of personnel or equipment on or below the surface does not create a right of ownership (Article 11).

It states that scientific research, exploration and use of the Moon must be free for all parties without discrimination of any kind (Article 6).

It hopes that samples obtained during research activities will be made available to all countries and scientific communities for research (Article 6.2).

It asks that all areas or regions that are found to have particular scientific interest be designated as international scientific reserves (Article 7.3).

It asks to promptly notify the United Nations and public opinion of any phenomenon that may endanger human life or health, as well as any indication of extraterrestrial life (Article 5.3).

The Moon Treaty reiterates most of the provisions already outlined in the OST and adds two new concepts to address the exploitation of natural resources in outer space: apply the concept of "common heritage of mankind" to outer space activities and ensure that participating countries develop a regime that establishes the appropriate procedures for mining. Two additions that seemed limiting to the great space powers that have not ratified it.

In fact, only 18 states have signed it and this makes the treaty on the use of the Moon a missed opportunity for space law, which continues to refer only to the OST.

National Laws

Space Launch Competitiveness Act

In order to increase private sector involvement in space, the Obama administration in 2015 introduced the U.S. Commercial Space Launch Competitiveness Act to provide a legislative framework for the mining of celestial bodies by private industries. The law refers to Article VI of the OST which explicitly mentions "the activities of non-governmental entities in outer space, including the Moon and other celestial bodies" stating that these "will require authorization and ongoing supervision by the appropriate contracting State" and that contracting States will have international responsibility for national space activities carried out by governmental or non-governmental entities.

The Space Launch Competitiveness Act explicitly allows U.S. citizens and industries to "engage in the exploration and commercial exploitation of space resources", including water and minerals.

To avoid conflict with the OST, it is clearly stated that "the United States does not assert [with this law] sovereignty or sovereign or exclusive rights or jurisdiction over, or ownership of, any celestial body", but rather offers legislative coverage to those who wish to exploit space resources. In other words, a mine on the Moon or an asteroid can be exploited by a private entity (for profit) without the nation that authorized this activity claiming property rights. All this must be done in compliance with Article IX, i.e., avoiding "potentially harmful interference with the activities of exploration and peaceful use of outer space, including the Moon and other celestial bodies" by other signatories of the OST. A delicate point, given that different entities, perhaps of different nationalities, might want to exploit the same asteroid or the same region of the Moon.

Following the American initiative, similar national laws to legalize the appropriation of extraterrestrial resources have been introduced by other countries, first and foremost **Luxembourg**, with the clear aim of attracting industries potentially interested in space mining, followed by **Japan**, **China**, **India**, and **Russia**.

Italy is also adopting a law to regulate access to space by operators active in the country. It is a law that looks at the development of the *space economy* trying to regulate the launch, release, in-orbit management, and return of space objects also for commercial suborbital activity as well as for any other activity carried out in outer space and on celestial bodies. Industry operators will have to ask for authorization to participate in space activities and the task of control will be entrusted to the Italian Space Agency.

As already mentioned, the European Commission is planning to adopt, in coordination with its member states, a Space Law aimed at improving the sustainability of the entire space supply chain to reduce pollution and improve orbital management.

The Artemis Accords

Unlike the OST treaty and subsequent conventions, the Artemis Accords are not emanations of United Nations offices but are rather co-managed agreements by NASA and the U.S. Department of State aimed at establishing a series of common principles to ensure that missions falling under the Artemis program are undertaken responsibly.

The Artemis Accords were launched on October 13, 2020 when they were signed by Australia, Canada, Italy, Japan, Luxembourg, United Arab Emirates, England, and the United States. Other states have signed the accords, which, as of mid-January 2025, count 53 signatories. These are all the states that want to make a large or small contribution to the Artemis program, which, under the guidance of NASA, aims to usher in a new era of space exploration and to bring the first woman and the first person of color to the Moon in 2027.

At the ceremony NASA Administrator Jim Bridenstine said "Fundamentally, the accords are about avoiding conflict, transparency, public registration, deconflicting activities. These are the principles that will preserve peace."

The Artemis Accords are a series of declarations that establish common principles and guidelines to be applied to the safe exploration of the Moon and, eventually, beyond, and their formulation fully respects both the original OST and subsequent agreements and conventions.

In more detail, the Artemis Accords cover:

Peaceful exploration of space: The nations agree that all activities conducted under the Artemis program must be carried out for peaceful purposes in accordance with international law.

Transparency: The signatory nations must conduct their activities transparently, in the hope of avoiding confusion and conflict. This also extends to signatories sharing scientific information with the public and the international scientific community in good faith. Signatories must also apply this principle to competing projects and are required to coordinate the publication of research and documents among themselves. The agreements state that: "The signatories of the Artemis Accords commit to making scientific information public, allowing the whole world to join us on the Artemis journey".

Interoperability: The agreements affirm that the nations participating in the Artemis program should aim to develop and provide support for systems that can operate together with existing infrastructures, hoping to improve both the safety of space operations and the sustainability of these missions.

Emergency assistance: The nations signing the Artemis Accords commit to assisting astronauts and personnel in outer space who are in difficulty.

Registration of space objects: The nations participating in the Artemis program must establish who among them must register any relevant space object.

Preservation of historical heritage: The signatories of the Artemis Accords have committed to preserving humanity's space heritage. This includes sites of historical importance such as human or robotic landing sites, artifacts, spacecraft, and other evidence of activity on other celestial bodies.

Space resources: The signatories of the agreement affirm that the extraction and use of space resources from the above-mentioned celestial bodies is fundamental to supporting safe and sustainable space exploration. They also commit to informing the Secretary-General of the United Nations, the public, and the scientific community about space resource extraction activities.

Prevention of interference: The nations signing the Artemis Accords commit to preventing harmful interference. This also concerns the establishment of so-called "safety zones", with areas that can be established with agreement between countries and that can be removed when the relevant operations cease.

Orbital debris: The countries signing the Artemis Accords commit to planning the safe, timely, and efficient disposal of debris as part of the mission planning process. The signatories of the agreements also agree on the need to limit the generation of new long-life or harmful debris. This includes the safe disposal of space structures in the post-operational phase of missions.